Planning Report 13-2

# Economic Case Study: The Impact of NSTIC on the Internal Revenue Service

Prepared by:
RTI International
for

National Institute of
Standards & Technology

July 2013

**National Institute of
Standards and Technology**
U.S. Department of Commerce

# Contents

**Appendixes**

# Figures

# Tables

# ACKNOWLEDGEMENTS

This study benefited from the contributions of many individuals and organizations. The authors wish to thank the many staff at the Internal Revenue Service who provided information on current IRS activities and the likely impact of new authentication mechanisms. The study also would not have been possible without the thoughtful contributions of identity providers and cyber security experts whose participation, while confidential, offered the necessary insight into how the identity ecosystem is expected to function under the National Strategy for Trusted Identities in Cyber Space. Lastly, the authors wish to thank several individuals in particular for their comments on draft material and their insights and recommendations over the course of the study. Many thanks to Jeremy Grant (NIST), Mike Garcia (NIST), Rich Phillips (IRS), Gregory Tassey (NIST), and Gary Anderson (NIST).

# LIST OF ABBREVIATIONS

| | |
|---|---|
| AGI | adjusted gross income |
| BAE | Backend Attribute Exchange |
| BCR | benefit-to-cost ratio |
| CA | Computer Associates |
| CMS | Centers for Medicare & Medicaid Services |
| CSIRC | Computer Security Incident Response Center |
| FCCX | Federal Cloud Credential Exchange |
| FICAM | Federal Identity, Credential, and Access Management |
| GSA | General Services Administration |
| HSPD-12 | Homeland Security Presidential Directive 12 |
| IDPs | identity providers |
| IP | identity protection |
| IRS | Internal Revenue Service |
| IT | information technology |
| ITIN | Individual Taxpayer Identification Number |
| LOA | level of assurance |
| NIH | National Institutes of Health |
| NIST | National Institute of Standards and Technology |
| NPE | nonperson entity |
| NPO | National Program Office |
| NPV | net present value |
| NSTIC | National Strategy for Trusted Identities in Cyberspace |
| O&M | operation and maintenance |
| OLS | Online Services |
| OMB | Office of Management and Budget |
| PII | personally identifiable information |
| PIN | personal identification number |
| PIV | Personal Identity Verification |
| PIV-I | PIV-interoperable |
| PKI | Public Key Infrastructure |
| RP | relying party |

| | |
|---|---|
| RUP | Registered Users Portal |
| SSA | Social Security Administration |
| SSN | Social Security Number |
| TAC | Taxpayer Assistance Center |
| TFP | trust framework provider |
| TIN | Taxpayer Identification Number |
| USDA | U.S. Department of Agriculture |
| USPS | U.S. Postal Service |

# ABSTRACT

The National Strategy for Trusted Identities in Cyberspace (NSTIC) offers a vision of more secure, efficient, and cost-effective authentication through widespread use of robust third-party credentials standardized to a national strategy. If successful, internet users would be able to use the same credentials to access secure online services everyplace they go online, saving time and reducing privacy risks. Websites and online service providers would be able to leverage externally provided, verified credentials and attributes, ultimately resulting in cost savings based on economies of scale and more services being offered online as a result of greater security.

The NSTIC National Program Office and the Economic Analysis Office at the National Institute of Standards and Technology funded this economic analysis to better understand the benefit-cost implications of the NSTIC model for U.S. federal agencies. The focus of the study was a comparison of the costs and benefits of a new Internal Revenue Service (IRS) proprietary authentication system and an NSTIC-aligned authentication system using third-party credentials. Up-front costs and annual costs and benefits for each authentication system scenario were estimated relative to the status quo for three hypothetical demand levels—20%, 50%, and 70% of U.S. taxpayers.

The analysis results suggest significant net benefits if the IRS adopts either a proprietary authentication system or an NSTIC-aligned authentication system. Improved online identity management would result in a net benefit of $74 million to $305 million annually for the IRS, relative to continuing current practices. Savings would move to the higher end of the range as taxpayer adoption increased. Furthermore, should public adoption of NSTIC credentials reach the hypothetical levels, and if the IRS adopts the NSTIC approach of leveraging accredited third-party credentials, the up-front adoption costs would be $40 million to $111 million less than an IRS-only proprietary authentication system. Ongoing annual costs would be $2 million to $19 million (10% to 52%) less, and the annual benefits would be $0.6 million to $1.9 million higher.

The additional benefits of adopting the NSTIC-aligned system result from avoiding having to pay to identity proof all existing and new taxpayers (users) and providing customer service and troubleshooting support to all users. In other words, the IRS would benefit from economies of scale in identity proofing and credential management.

The IRS incurs significant costs as a result of identity fraud—likely over $5 billion per year—much of which occurs through e-filing. This study does not claim any net effect of improved authentication on fraud. Improved authentication may result in a reduction in online identity theft, though identity theft associated with paper returns may increase because it will be easier to commit fraud on paper relative to electronic returns. Nonetheless, the IRS could focus its attention on tax submissions that have not been authenticated, and the public would have a safe way to conduct transactions with the IRS with minimal risk of identity theft.

For any public and private sector organization considering NSTIC adoption, the methodology and results of this case study provide a guide for the types of costs and benefits that should be considered.

# EXECUTIVE SUMMARY

Authentication of an internet user's true identity is becoming even more important as the number of sensitive online services and the types of electronic communication increase. In 2007, the typical computer user had about 25 password-based user accounts (Florencio & Herley, 2007), and that number has likely risen. Because they have so many accounts, users often create simple passwords to increase their ability to remember them, reuse usernames and passwords on multiple websites, and write down usernames and passwords to prevent loss. These habits result in a reduction in security.[1]

Not surprisingly, consumers are pushing back against website requests to create new accounts. A study by Blue Research (2011) found that 77% of internet users change their behavior when faced with online account registration—about 60% leave a site if presented with a registration prompt, and 17% go to a different site if possible. Beyond being frustrating to internet users, this situation also represents a loss of business for companies.

According to Verizon's 2012 Data Breach Report, 44% of breaches in 2011 resulted from the use of easily guessable login credentials, 32% from the use of stolen credentials, and 5% from insufficient authentication (e.g., no login required) (Verizon, 2012).[2] At the same time, consumers are increasing the amount of shopping they do online: between 2011 and 2012 the share of retail shopping that takes place online increased by 11% (Census 2012). Together, these trends raise significant concerns about the risk of identity theft which, in 2011, is estimated to have affected over 8 million Americans and cost over $30 billion (Javelin, 2011).

Private and public sector organizations are spending billions of dollars trying to prevent unauthorized access to their IT systems and to mitigate the damage when unauthenticated access occurs. This is in addition to the opportunity costs of not putting services online because of security risks and unacceptably high authentication costs.[3]

## ES.1 National Strategy for Trusted Identities in Cyberspace

The National Strategy for Trusted Identities in Cyberspace (NSTIC) offers a vision of more secure, efficient, and cost effective authentication through standardizing and expanding the use of third-party credentials. Today, many internet users use their login credentials—username and password—from their Google, Facebook, Yahoo, and other accounts to log into other

---

[1] Simple passwords are easier for hackers to guess. Reusing usernames and passwords can result in multiple online accounts being compromised when the credentials used for one website are stolen. Writing down usernames and passwords results in thieves being able to acquire passwords visually.

[2] Note that these percentages are potentially overlapping so should not be added together. Furthermore, larger organizations tend not to have as many breaches resulting from inadequate authentication as smaller organizations.

[3] According to research by the Corporate Executive Board, financial companies spent approximately $39 billion on authentication and identity management solutions in 2012 worldwide (Malo, 2012).

websites. However, these credentials lack standardized security and privacy policies and procedures dictating how and what information is shared between identity providers and websites that accept the credentials they issue. In addition, secure third-party credentials of the type needed for sensitive online transactions are not readily available.

Published by the White House in April 2011, the NSTIC outlines a strategy to

- develop a comprehensive identity ecosystem framework, build and implement interoperable identity solutions,

- enhance confidence and willingness to participate in the identity ecosystem, and

- ensure the long-term success and viability of the identity ecosystem (White House, 2011).

The NSTIC proposes that organizations who are expert in identity verification, credential management, and authentication administer the authentication process. These organizations, called identity providers or IDPs, would authenticate individual users and provide them with credentials. Other organizations, called relying parties or RPs, would then accept these third-party credentials from their employees, business partners, and customers seeking to conduct online communications and transactions.

Relying parties could choose from several categories of credentials to accept, differentiated by (1) the level of assurance they provide (e.g., how rigorously they verify individuals' identities) and (2) the strength of the credentials issued (e.g., is multifactor authentication used).[4] Users would then acquire credentials based on the requirements set by the websites they aim to access.

Users could choose to use as few as one set of credentials to access multiple websites, saving them time and reducing the vulnerability of their personal information to theft. Websites and service providers would be able to leverage economies of scale by not having to proof the identity of each user of their system and may place more services online based on the increased security of the authentication system.

---

[4] The NSTIC proposes that credentials be offered with varying levels of assurance (LOAs). Although these levels are not specified in the Strategy, OMB Memorandum 04-04 (Bolton, 2003) provides an example of such a system. Briefly, LOA-1 requires only that an individual have a username and password, LOA-2 builds on LOA-1 by requiring that the individual's identity be verified or "proofed" through an approved process, LOA-3 builds on LOA-2 by requiring two-factor authentication (e.g., a username and password plus a generic token that shows a new personal identification number [PIN] every minute), and LOA-4 builds on LOA-3 by requiring a hardware cryptographic token and in-person identity proofing.

## ES.2 Case Study Comparison of Proprietary versus NSTIC-Aligned Authentication at the IRS

In order to assess the potential net benefits of NSTIC implementation, the NSTIC National Program Office (NPO) and Economic Analysis Office at the National Institute of Standards and Technology (NIST) funded an economic analysis of the potential identity management solutions at the IRS. RTI International conducted this study between October 2011 and July 2013.

The focus of the study was a comparison of the costs and benefits of a proprietary authentication system to an NSTIC-aligned one. Currently, the IRS is developing a pilot implementation of a proprietary authentication system that would include identity proofing up to level of assurance (LOA)-2 and ongoing management of credentials and authentication. The IRS is also considering accepting LOA-2 and LOA-3 third-party credentials in the future, but no timeline has been identified. This study calculated the costs and benefits of these two scenarios as compared with the status quo for three hypothetical levels of demand: 20%, 50%, and 70% of U.S. taxpayers.[5]

The proprietary system will require the IRS to pay for identity proofing for all taxpayers who use the system, whereas accepting NSTIC-aligned third-party credentials will likely result in those costs being borne by IDPs and then recouped via per-authentication transaction fees paid by the IRS to IDPs. The up-front costs of implementation—labor, capital, and services needed to develop the infrastructure for new credentials to be accepted—will be similar for the two scenarios, except for the identity proofing costs. Table ES-1 lists the primary types of costs that will be incurred up-front and on an ongoing basis for each of the two future authentication scenarios.

In both cases, improving the security of online transactions would enable the IRS to increase its online service offerings and may help fight identity fraud. If taxpayers were able to use new IRS online services for communicating with the IRS, the IRS could save money relative to providing the same services by phone, by paper, and in person, and taxpayers would benefit from faster and higher quality service. Table ES-2 provides a list of the potential types of benefits that would result from improved authentication.

---

[5] Of note, in addition to these two scenarios, a third option exists. The Federal Cloud Credential Exchange (FCCX) is a government initiative that aims to allow federal agencies to use a single intermediary to verify the identities of their customers. Like the NSTIC, this initiative aligns focuses on the government using third-party credentials. The FCCX might offer agencies additional cost savings over and above those likely to result from an IRS proprietary authentication system, as costs are distributed over numerous agencies and economies of scales are realized.

**Table ES-1.   IRS Primary Cost Categories, by Future Authentication Scenario**

| Type of Cost | Activity | IRS Proprietary Authentication System | NSTIC-Aligned Authentication System |
|---|---|:---:|:---:|
| Up-front fixed costs | ▪ Updating the core infrastructure to accept new authentication credentials | ● | ● |
| | ▪ Updating infrastructure to enable authenticated application use (per new application) | ● | ● |
| | ▪ Establishing a relationship with third-party IDPs | ● | ● |
| | ▪ Conducting identity proofing for each taxpayer | ● | ○ |
| Ongoing/ variable costs | ▪ Maintaining the core infrastructure and carrying out any changes necessary | ● | ● |
| | ▪ Maintaining the application-specific authentication infrastructure | ● | ● |
| | ▪ Providing customer support (e.g., reissuance of credentials) | ● | ● |
| | ▪ Conducting identity proofing for new taxpayers each year | ● | ○ |
| | ▪ Verifying credentials for each transaction session | ● | ● |

● = relevant cost category. ○ = not relevant cost category.

To create a conceptual model for the analysis and later to collect data needed for estimating the economic impact metrics, RTI interviewed more than 40 individuals, including IRS staff, IDPs, attribute providers, other government agencies, and identity experts. A variety of publically available documents and data and many IRS internal documents were also reviewed as part of the research.

## ES.3  Economic Impact Estimates

While adopting either the propriety authentication system or an NSTIC-aligned system would result in significant net benefits ranging from $74 million to $305 million annually, both the projected costs and the projected benefits would differ significantly between the two systems. The analysis results suggest that if the IRS adopts the NSTIC, the up-front, one-time costs will be $40 million to $111 million less than if the IRS decides to adopt a new proprietary authentication system.

The projected annual net benefits if the IRS adopts a proprietary system are $74 million (20% of U.S. taxpayers), $203 million (50%), and $286 million (70%), whereas with an NSTIC-

**Table ES-2.  IRS Benefit Categories Resulting from Improved Authentication**

| Area of Potential Benefit | Description |
|---|---|
| *Benefits Quantified in this Study* | |
| Efficiency of online services | • Reduction in call and mail volume for services that could be provided automatically online<br>  – Phone calls from taxpayers requesting account information and products (such as transcripts) to be mailed back<br>  – Letters received from taxpayers for various account inquiries or requests<br>  – Letters and notices related to account inquiries or special services (not return related) mailed to taxpayers |
| Reduced authentication-related customer service activities | • Reduction in phone assistor support in dealing with<br>  – Login issues to the RUP<br>  – E-file PIN requests |
| *Benefits Discussed Qualitatively in this Study* | |
| Increased e-filing | • Reduction in paper return processing costs<br>  – Costs incurred prior to transcription into electronic format<br>  – Costs of corresponding with taxpayers by mail as a result of errors |
| Identity theft reduction | • Reduction in labor costs of mitigating cases of identity theft (IRS could focus attention on taxpayers who do not use NSTIC credentials)<br>  – Cost of communicating by phone, by mail, and in person<br>  – Criminal investigation of identity theft cases<br>• Reduction in capital costs of phone calls and mail items |

aligned system, the IRS should expect annual benefits of $76 million (20% demand), $208 million (50%), and $305 million (70%).

The differences in the annual net benefits are because of the existence of identity proofing costs only under the proprietary system and because of decreasing transaction fees for higher levels of demand under the NSTIC-aligned system. Annual operations and maintenance costs for IT infrastructure are expected to be comparable in the two scenarios.

Although the annual net benefits are only slightly lower in the proprietary system, the difference between the IRS proprietary system and an NSTIC-aligned authentication system can be seen most visibly in the difference in up-front adoption costs. Table ES-3 breaks out the costs and benefits projected for each scenario.

## ES.4  Conclusions

The NSTIC offers a vision through which online authentication can be made more secure and less expensive, resulting in significant time and monetary savings for organizations and individuals. This study of the net benefits to the IRS of NSTIC adoption suggest that if demand for new credentials reaches 20% of taxpayers or higher, the annual net benefits will be

**Table ES-3.** One-Time Costs and Annual Costs and Benefits, by Scenario and Level of Adoption ($000)

| Scenario | NSTIC Demand | | | | | |
|---|---|---|---|---|---|---|
| | 20% | | 50% | | 70% | |
| | One-Time Costs | Annual Costs and Benefits | One-Time Costs | Annual Costs and Benefits | One-Time Costs | Annual Costs and Benefits |
| **Scenario 1: Base Case** | $— | $— | $— | $— | $— | $— |
| **Scenario 2: IRS Proprietary Authentication System** | | | | | | |
| Benefits | | $90,942 | | $227,354 | | $318,295 |
| Total one-time costs | $68,816 | | $118,764 | | $139,445 | |
| *One-time adoption (infrastructure) costs* | *$28,230* | | *$28,230* | | *$28,230* | |
| *One-time identity-proofing costs* | *$40,586* | | *$90,534* | | *$111,215* | |
| Annual costs | | $16,856 | | $24,801 | | $31,841 |
| Annual net benefits | | $74,086 | | $202,553 | | $286,454 |
| **Scenario 3: NSTIC-Aligned Authentication System** | | | | | | |
| Benefits | | $91,498 | | $228,744 | | $320,242 |
| Total one-time costs | $28,230 | | $28,230 | | $28,230 | |
| *One-time adoption infrastructure) costs* | *$28,230* | | *$28,230* | | *$28,230* | |
| *One-time identity-proofing costs* | *$—* | | *$—* | | *$—* | |
| Annual costs | | $15,126 | | $20,942 | | $15,126 |
| Annual net benefits | | $76,372 | | $207,802 | | $305,116 |
| **Comparison of Scenario 3 with 2** | | | | | | |
| **One-time cost comparison (Cost savings of Scenario 3)** | $40,586 | | $90,534 | | $111,215 | |
| **Annual net benefits comparison (Cost savings of Scenario 3)** | | $2,286 | | $5,249 | | $18,662 |

significant. Compared with a proprietary authentication approach, the up-front costs will likely be much lower to adopt the NSTIC, and the annual net benefits of an NSTIC-aligned system will be higher. Moreover, if an NSTIC-based system results in a higher level of adoption because the credentials can be used at other websites, then the annual net benefits to the IRS could be much greater in the NSTIC scenario. While there is still uncertainty around the user adoption rate of NSTIC-aligned solutions, if these hypothetical levels of adoption are met, significant net benefits will result.

Additional work by the NSTIC NPO and the greater identity ecosystem community will ensure that the assumptions made in this report are accurate. For other organizations, public and private, considering NSTIC adoption, the results of the IRS economic impact case study analysis and the methodology presented in this report provide a guide for the types of costs and benefits that should be reviewed.

# 1. INTRODUCTION

The growth in identity theft—stealing someone's identity—and identity fraud—using that stolen identity for illicit gain—increases the cost and risk of conducting sensitive transactions online. For some organizations, such as financial institutions, the benefits of conducting certain transactions online outweigh the significant costs of securing those transactions. For others the costs and risks are deemed too great to bear. Similarly, customers have legitimate concerns about the security of their personal information, which they are asked to provide to multiple organizations for online user account management.

Today, someone who connects online to multiple service providers (e.g., a bank, credit card company, cable company, telephone company, and hospital) may create a new user account for each provider, often reusing usernames and passwords. Each provider collects and stores sensitive information about the individual to authenticate his or her identity. Having personally identifiable information replicated over and over again with different service providers compounds the vulnerability of one's personal information to theft.

In April 2011, the White House released the National Strategy for Trusted Identities in Cyberspace (NSTIC), which called for the creation of an "identity ecosystem" that would enable individuals to choose among multiple identity providers (IDPs) and digital credentials to be used to conduct more convenient, secure, and privacy-enhancing transactions everyplace they go online. The purpose of the NSTIC is to create an online environment where individuals and organizations can trust each other through the use of a common, cost-effective means of asserting identity at various levels of assurance.

The NSTIC vision of third-party credentialing would establish a user account authentication paradigm where one IDP issues credentials[1] to a user, and these credentials can be used for multiple online websites, called relying parties (RPs). Third-party credentialing exists today; many websites allow users to log in using their Yahoo!, Google, or Facebook accounts, for example.

The NSTIC aims to elevate current instantiations of third-party credentialing to a higher level of security, standardize third-party authentication processes, and increase adoption. IDPs would proof and validate the identity of users, in some cases using methods that provide a high level of assurance, and then issue users credentials that could be trusted by any approved RP. The advantage for websites that accept third-party credentials is that they could trust IDPs and would not need to conduct costly proofing, validation, and user account maintenance on their

---

[1] An example of a credential may be a username and a password, as well as another "factor," such as a one-time password delivered via text message or through a mobile phone application, which would be needed for more secure transactions.

own. Rather, they would pay a small transaction charge, similar to the way payment card issuers charge a merchant for accepting a credit card, to the IDP. Ultimately, the NSTIC aims to lower costs, lower risks, and increase the security of online transactions.

This report presents a case study analysis of the potential benefits to the Internal Revenue Service (IRS) of accepting NSTIC-aligned third-party credentials to provide secure online services to citizens. Most citizens only interact with the IRS during the tax preparation season, but when they do, they exchange such sensitive information as their social security number, income, and address. The IRS plans to increase the services available to citizens online, but to do so it must first determine how to sufficiently authenticate users.

In 2012, the IRS spent over $1 billion communicating with taxpayers on paper and by telephone. It processed 30 million paper-based tax returns, despite the fact that electronic filing (e-filing) is available to taxpayers. These service-related costs could be significantly reduced if many of the services provided by the IRS were shifted online and more citizens used existing online services.

Identity theft that results in tax fraud imposes additional costs on the IRS, particularly the losses (money paid out to individuals committing tax fraud) that are not recovered and the labor costs incurred to help prevent and mitigate cases of identity theft. In 2011, the IRS Inspector General estimated the unrecovered losses resulting from identity theft-related tax fraud to be more than $5 billion (George, 2012). Implementation of the NSTIC may not reduce fraudulent returns because criminals could switch to submitting fraudulent tax returns on paper instead of electronically. Nonetheless, if electronic filers (e-filers) began using third-party credentials, this would enable the IRS to concentrate its identity theft and fraud teams' attention on paper returns.

The adoption of third-party credentials, as proposed by the NSTIC, offers an approach that could help achieve the IRS's aims. By increasing the level of assurance (LOA) in online transactions, NSTIC-aligned third-party credentials could enable the IRS to conduct more transactions with taxpayers online in a secure setting while lowering the cost of user identity management.

This case study explores the costs and benefits of the NSTIC through the microcosm of the IRS, serving as an example of the costs and benefits that other federal government agencies and private-sector organizations may accrue should they decide to adopt the NSTIC and accept third-party credentials. The NSTIC National Program Office (NPO) and Economic Analysis Office at the National Institute of Standards and Technology (NIST) are sponsoring this independent prospective economic analysis to enhance understanding of the cost-savings potential of the NSTIC.

## 2. THE NSTIC VISION OF AN ONLINE IDENTITY ECOSYSTEM

The NSTIC proposes that the private sector provide third-party credentials, which are bound to an online identity ecosystem by a common framework, established by a private sector-led steering group consisting of both public and private sector stakeholders. The NSTIC proposes a system in which strong authentication credentials can be granted by one organization and accepted by many others. The strategy fights identity theft and fraud while simultaneously reducing the social cost of identity management and individuals' need for separate authentication credentials for every website.

Broadly, the NSTIC has four major goals (The White House, 2011, p. 29):

- Develop a comprehensive identity ecosystem framework.

- Build and implement interoperable identity solutions.

- Enhance confidence and willingness to participate in the identity ecosystem.

- Ensure the long-term success and viability of the identity ecosystem.

The identity ecosystem would enable both private and public sector organizations to improve the authentication of business-to-business, business-to-consumer, and consumer-to-business interactions. In such a system, rigorous identity proofing assures authentication. Over the long term, full implementation should result in higher levels of trust in sensitive online interactions, a higher volume of transactions conducted online, and a reduction in the costs to both firms and individuals of authenticating users and maintaining credentials.

The role of the public sector is limited to coordinating activities, providing technical infrastructure, promoting information sharing among stakeholders, and seeding the marketplace through grant funding. Technical infrastructure such as standards and standard protocols will play a critical role in ensuring interoperability and ensuring that levels of assurance of third-party credentials provided by an IDP are standardized and can be verified and certified. To this end, the NSTIC NPO at NIST is playing a role in catalyzing the development of the technical infrastructure necessary for the NSTIC to be implemented efficiently and effectively.[1]

The federal government is also creating demand for NSTIC solutions through its implementation of the Federal Identity, Credential, and Access Management (FICAM) Roadmap, a broad government initiative driven by the Federal CIO Council that aims to standardize the transmission and storage of digital credentials for online communication across

---

[1] NIST has supported authentication for many years. The Cyber Security Division within NIST's Information Technology Laboratory (ITL) has developed a variety of standards, standard procedures, best practice documents, and other technology infrastructure components that support online authentication. Many of these components, developed as far back as 15 to 20 years, are still affecting current authentication activities today.

the federal government (2011). Although FICAM primarily provides guidance for adopting and using internal credentials (i.e., Personal Identity Verification [PIV] cards), which are required for all federal government agencies, FICAM also recommends that agencies build relationships with IDPs or intermediary parties to enable the acceptance of nongovernment credentials. Additional information on FICAM is provided in Appendix A.

## 2.1 Levels of Assurance in Authentication

Authentication in the online world is synonymous with "trust" in the offline world. Similar to the need for physical banks to trust that a customer talking to a teller is who he says he is, public and private organizations conducting high-risk transactions online need to be able to ensure that they know with whom they are communicating. In many cases, online transactions require minimal authentication. Establishing a user account with many websites can be done with only an e-mail address so that the sites know that the same user is returning to a specific account. Yet a bank offering online customer account access and modification needs to have a high level of trust that the person who logs in is the person who owns the account. In both cases, trust is important, but the level of trust and the associated authentication costs differ significantly.

The value proposition offered by many online products and services is strongly tied to the ability to authenticate users. From a business standpoint, when authentication is broken, the quality of the business' products or services, as viewed by customers or users, may be diminished;[2] thus, customer adoption, use, and retention may be affected, decreasing related revenues.

Authentication is primarily determined by the presentation and acceptance of a user's credentials—a set of factors, such as a username and password—that are used by a host or system to verify the user's identity. A credential's LOA determines how reliably the credential can assert the identity of an individual.

There are four distinct LOAs, as defined by the Office of Management and Budget (OMB) in *E-Authentication Guidance for Federal Agencies* (Bolten, 2003), and for which technical specifications are provided by NIST in Special Publication 800-63-1 *Electronic Authentication Guideline* (Burr et al., 2011).[3] Each of the four levels is determined based on the requirements of four criteria.

**The first criterion is the requirement of a token**. NIST defines a token as "something that the claimant possesses and controls (typically a cryptographic module or password) that may

---

[2] Authentication of various types is used by many online companies to personalize users' experiences.
[3] This special publication supersedes NIST Special Publication 800-63 *Electronic Authentication Guideline*, which was released in 2006 (Burr, Dodson, & Polk, 2006).

be used to authenticate the claimant's identity" (Burr et al., 2011 , p. 15). NIST identified three factors as the cornerstones of authentication (Burr et al., 2011, p. 20):

- something you know (e.g., a password)

- something you have (e.g., an ID badge or a cryptographic key)

- something you are (e.g., a fingerprint or other biometric data)

While a token is often a password or an ID badge, different types of tokens or token combinations may be required depending on the level of security desired.

**The second criterion is identity proofing**. Identity proofing can involve the use of, for example, a government-issued ID (e.g., a driver's license), a financial account number, or past addresses. For higher levels of assurance, in-person appearance and verification of documents may be required.

**The third criterion is the set of authentication protocols** and the tokens and other credentials that they entail, altogether known as the remote authentication mechanism. An authentication protocol is a set of rules for communication between a user with a token and the identity verifier to prove that the user is the owner of the credential/identity.

**The fourth criterion is the assertion protocol**, which dictates how identity verifiers communicate to an RP that an identity has been proofed. An assertion is generally transmitted through a secure message, or it may be presented as an object that is "digitally signed," which means that it is trusted to assert a user's identity. The assertion protocol is the same for all LOAs.

Each LOA has different requirements regarding the first three above criteria, as summarized in Table 2-1. The four LOAs build on each other and, at a high level, can be described as follows:

- **LOA-1** requires that an individual have a username and password.

- **LOA-2** builds on LOA-1 by requiring that the individual's identity be verified or "proofed" through an approved process.

- **LOA-3** builds on LOA-2 by requiring two-factor authentication (e.g., a username and password plus a generic token that shows a new personal identification number [PIN] every minute).

- **LOA-4** builds on LOA-3 by requiring a hardware cryptographic token and in-person identity proofing.

**Table 2-1.  Summary Definitions of NIST-Defined Levels of Assurance**

| LOA | Tokens | Identity Proofing | | Remote Authentication Mechanism |
| --- | --- | --- | --- | --- |
| | | Requirement Documentation | Required Procedures | |
| LOA-1 | Plain-text password transmission is not allowed, but cryptographic methods are not required. | None | None | Successful authentication requires that the claimant prove through a secure authentication protocol that he or she controls the token. |
| LOA-2 | Allows any of the token methods of LOA-3 or LOA-4, as well as passwords and PINs. | A valid current government ID (e.g., a driver's license or passport) number and a financial account number (e.g., checking account, savings account, loan or credit card, or tax ID). | Confirms information provided via records—including either the government ID or the account number. Confirms that name, date of birth, address, and other personal information in records are *on balance* consistent with the application and sufficient to identify a unique individual. | Successful authentication requires that the claimant prove through a secure authentication protocol that he or she controls the token. |
| LOA-3 | Requires cryptographic strength mechanisms that protect the primary authentication token (secret key, private key, or one-time password); multifactor authentication is required. | A valid government ID (e.g., a driver's license or passport) number and a financial account number (e.g., checking account, savings account, loan, or credit card, or tax ID) with confirmation via records of both numbers. | Confirms information provided via records—including *both* the government ID and the account number. Confirms that name, date of birth, address, and other personal information in records are consistent with the application and sufficient to identify a unique individual. | Successful authentication requires that the claimant prove through a secure authentication protocol that he or she controls the token, and the claimant must first unlock the token with a password or biometric or must also use a password in a secure authentication protocol, to establish two-factor authentication. |
| LOA-4 | Requires a hardware cryptographic token validated at FIPS 140-2 Level 2 or higher overall with at least FIPS 140-2 Level 3 physical security. | In-person appearance and verification of two independent ID documents or accounts, meeting the requirements of Level 3. | Confirms information provided via records—including both the government ID and the account number. Compares picture(s) with applicant. Confirms that name, date of birth, address, and other personal information in records are consistent with the application and sufficient to identify a unique individual. | Authentication requires that the claimant prove through a secure authentication protocol that he or she controls the token with a password or biometric or must also use a password in a secure authentication protocol, to establish two-factor authentication. |

Source: Burr et al. (2011).

Federal government agencies, such as the IRS, with plans to transmit individuals' personally identifiable information (PII) are likely to require LOA-3 credentials for many types of applications, though LOA-2 credentials should be sufficient for some applications. Many bank authentication systems have recently been moving from requiring LOA-2 credentials to requiring LOA-3 credentials. The remainder of the discussion in this report focuses on LOA-3 credentials, which were used for the case study analysis described in Chapter 4.

## 2.2 A Third-Party Identity Ecosystem Under the NSTIC

The NSTIC's vision of an identity ecosystem consists of six primary stakeholders and components responsible for providing and accepting authentication credentials:

- An **individual** is a person engaged in an online transaction. Individuals are the first priority of the Strategy.

- A **nonperson entity (NPE)** may also require authentication in the identity ecosystem. NPEs can be organizations, hardware, networks, software, or services and are treated much like individuals within the identity ecosystem. NPEs may engage in or support a transaction.

- The **subject** of a transaction may be an individual or an NPE.

- **Attributes** are a named quality or characteristic inherent in or ascribed to someone or something (e.g., "this individual's age is at least 21 years").

- A **digital identity** is a set of attributes that represent a subject in an online transaction.

- An **IDP** is responsible for establishing, maintaining, and securing the digital identity associated with that subject. These processes include revoking, suspending, and restoring the subject's digital identity if necessary (The White House, 2011).

Authenticated individuals and NPEs receive trusted credentials from IDPs[4] and access services from RPs that accept these credentials. The IDP can provide a credential to an individual. When the individual uses the credential to access a service (e.g., online banking) offered by an RP (e.g., the bank), the IDP will assert the individual's identity to the RP. The RP can trust the IDP because the framework will ensure that the IDP meets all of the commonly agreed upon criteria to provide a specific level of assurance. Figure 2-1 shows an example of credential issuance in the identity ecosystem.

---

[4] For government agencies to accept credentials from IDPs, the IDPs must be approved. Approval of IDPs requires two steps. First, identity trust framework providers (TFPs) must apply to the General Services Administration (GSA) for their policies and procedures to be deemed to have met LOA requirements as defined by NIST. Second, TFPs then serve as auditors in the process of IDP approval. Each IDP must apply to a TFP, and the TFP must approve that they have met the TFP's requirements. Private organizations looking to accept third-party credentials can also decide to only accept credentials issued by IDPs who have been approved through this process.

**Figure 2-1.    An Example of Credential Issuance in the Identity Ecosystem**

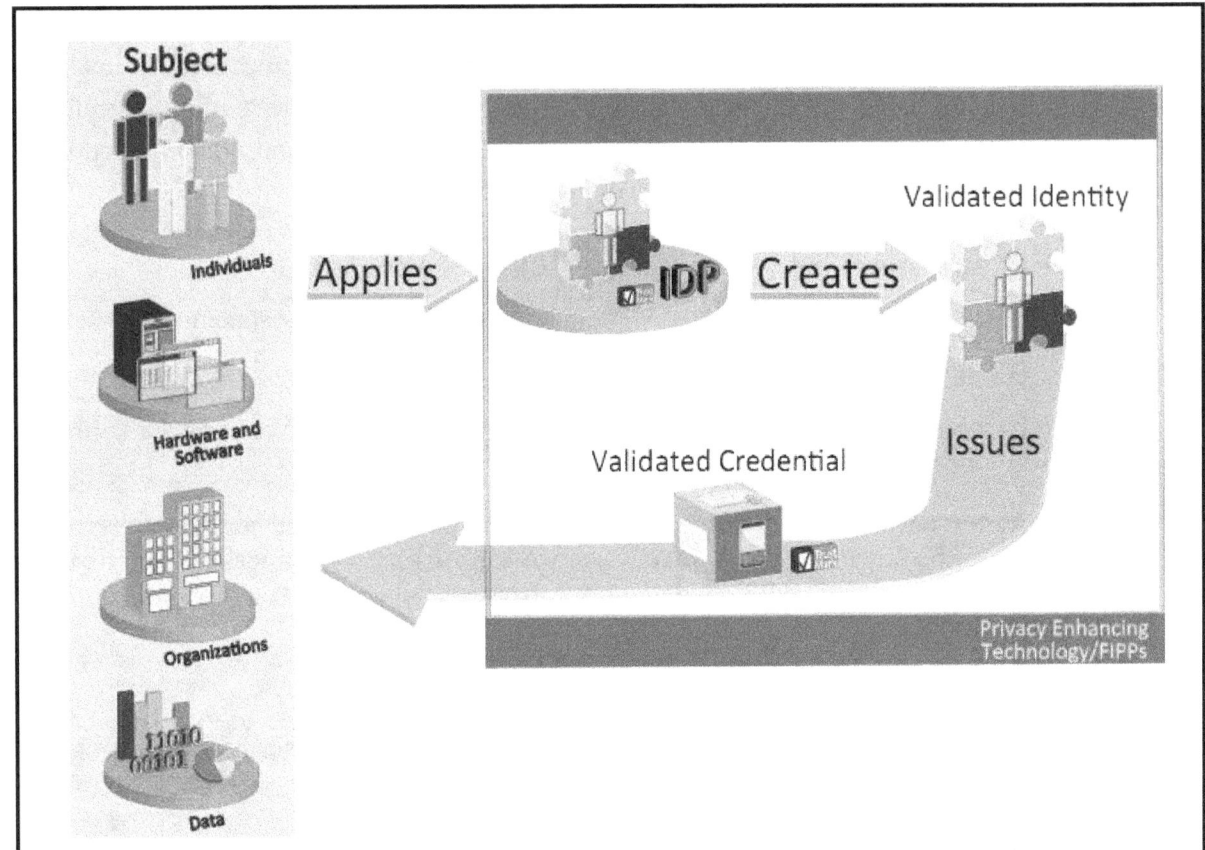

Source: The White House (2011).

IDPs' business model in the identity ecosystem envisioned by the NSTIC is based on widespread use of credentials. Assuming high levels of demand for third-party credentials by internet users and RPs, IDPs will be able to streamline costs for identity proofing, credential management, and authentication verification, and as such, can charge RPs low per-transactions prices (each time users log in to a website using IDPs' credentials). If demand is not as high, IDPs will charge higher per-transaction prices and may charge RPs and/or internet users for identity proofing and credential issuance and management.

## 2.3    Current Status of the NSTIC

Since the announcement of the NSTIC by the White House in mid-2011, the NSTIC NPO was established at NIST and Congress appropriated $16.5 million to the program office for the 2012 fiscal year (112th Congress, n.d.), primarily to help establish a privately-led Identity

Ecosystem Steering Group and fund a set of five pilot projects.[5] The NPO crafted recommendations for establishing the Steering Group after receiving extensive public input on the steering group initiation plan, structure, international coordination activities, and policy frameworks[6] (NIST, 2012a). In August 2012, the Identity Ecosystem Steering Group formally convened for the first time with nearly 1,000 participants.[7]

Although the standardized policies and procedures that will help guide widespread implementations of the NSTIC are still in development,[8] some components of the NSTIC vision are currently being implemented. As of January 2013, four organizations have been approved by the GSA as TFPs: InCommon, Kantara, Open Identity Exchange, and Safe/BioPharma.[9] Once a TFP meets certain requirements,[10] it is then able to approve organizations that apply to be IDPs. Additionally, as of this writing, TFPs have approved seven organizations as IDPs for LOA-1, LOA-2, or LOA-3[11] (see the list of organizations by approved LOA in Table 2-2).

In terms of RPs, currently, the National Institutes of Health (NIH), through its iTrust framework, operates the only authentication framework that fully complies with the NSTIC's goals and framework (NIH, 2010). However, widespread federal government implementations of the NSTIC are expected in the near future. A memorandum issued by the Federal Chief Information Officer in 2011 mandated the acceptance of third-party credentials for low-security transactions following the federal government's approval of at least one TFP (OMB, 2011).

---

[5] In February 2012, the NSTIC NPO announced a competition to award a total of $10 million for pilot projects in support of the NSTIC and held a bidders conference later in the month. Winners were announced in September 2012; see the news release at http://www.nist.gov/itl/nstic-092012.cfm.

[6] See the NPO's recommendations document at http://www.idecosystem.org/filedepot_download/282/293.

[7] In addition to the Steering Group meetings, other stakeholder outreach efforts have been held. NIST held a series of workshops to work with the private sector in developing ideas for implementing the NSTIC. In March 2012, NIST held the IDtrust Workshop discussing how standards and technologies can help implement the identity ecosystem. Topics included usability, trust frameworks, business models, and privacy-enhancing technologies. In May 2012, the White House hosted a colloquium, in which senior White House staff presented NSTIC objectives and held a panel with private-sector representatives; in addition to the panel, representatives from over 30 major corporations were interested in the identity ecosystem (NIST, 2012b). See a list of key NSTIC-related documents and meetings at http://www.idecosystem.org/page/nstic-timeline-and-documents.

[8] Of note, much work on developing standardized frameworks for transmitting information has already been done. For example, OASIS, a global consortium, supported the development of a common protocol called the Security Assertion Markup Language (SAML) for communicating user authentication and attribute information. The SAML framework "allows business entities to make assertions regarding the identity, attributes, and entitlements of a subject ... to other entities, such as a partner company or another enterprise application" (Wisniewski, Nadalin, Cantor, Hodges, and Mishra, 2005, p. 2).

[9] See details on each of the approved TFPs at http://www.idmanagement.gov/pages.cfm/page/ICAM-TrustFramework-Provider.

[10] See details on the GSA's requirements for each TFP to be approved at http://www.idmanagement.gov/documents/trustframeworkprovideradoptionprocess.pdf.

[11] Several additional providers are approved for PKI-based LOA-3 and LOA-4.

**Table 2-2.  Approved IDPs, LOA-1 through LOA-3, as of January 2013**

| IDPs | LOA-1 | LOA-2 | LOA-3 |
|------|:-----:|:-----:|:-----:|
| Equifax | ● | ○ | ○ |
| Google | ● | ○ | ○ |
| PayPal | ● | ○ | ○ |
| Symantec | ● | ● | ● |
| VeriSign | ● | ○ | ○ |
| Verizon | ● | ● | ● |
| Wave Systems | ● | ○ | ○ |

● = approved.  ○ = not approved.

Source: IDManagement.gov. As of January 2013. Approved Identity Providers.
   http://www.idmanagement.gov/pages.cfm/page/icam-trustframework-idp..

Given that four TFPs have been approved, all agencies' websites that are created, enhanced, or upgraded must be able to accept third-party credentials if the websites allow citizens or business partners to register or log on and require LOA-1 authentication (OMB, 2011). This requirement applies to websites with LOA-2 through LOA-4 authentication where resources allow.

## 2.4  Proprietary Authentication versus NSTIC-Aligned Authentication

For individual organizations and government agencies seeking to improve online authentication, they can either develop their own proprietary system or accept NSTIC-aligned third-party credentials. Table 2-3 provides a list of key advantages and disadvantages of each approach.

Three federal government agencies have developed or are developing proprietary electronic authentication solutions for citizens' or organizations' use:[12]

- **U.S. Department of Agriculture (USDA)**. The USDA offers LOA-1 and LOA-2 accounts for users. The agency requires an LOA-2 account for users "conducting official electronic business transactions via the Internet," "entering into a contract with the USDA," or "filling out and submitting electronic forms or applications for USDA via the Internet" (USDA, 2012). Registration for an LOA-2 account requires that an applicant (in addition to registration) visit a "local registration authority at a USDA Service Center" to validate his or her identity using photo identification. Although USDA allows a wider range of online interaction with citizens, enrollment for a sufficient level of security requires offline correspondence.

---

[12] Of note, these agencies may also decide to accept third-party credentials in the future. Developing and managing a proprietary authentication system does not prevent an organization from accepting third-party credentials.

**Table 2-3.    Summary of Perceived Advantages and Disadvantages: Proprietary Authentication Versus NSTIC-Aligned Authentication**

| | Proprietary Authentication System | NSTIC-Aligned Third-Party Authentication System |
|---|---|---|
| Advantages | ▪ Full control of security policies and procedure<br><br>▪ More control of costs/pricing models used<br><br>▪ Quicker adoption (because policies, procedures, and pricing models are not finalized for NSTIC-aligned third-party credential authentication) | ▪ No need/cost for infrastructure, license agreements, or service agreements for identity proofing, credential management, and authentication<br><br>▪ No need for specialized staff to manage full authentication system |
| Disadvantages | ▪ Potential need for/cost of additional infrastructure (up-front and operation and maintenance [O&M]), license agreements, or service agreements for identity proofing, credential management, and authentication<br><br>▪ Need for specialized staff (or outsourced staff) to maintain full authentication system | ▪ Lack of control of security policies and procedures (determined by IDPs within bounds of GSA guidelines)<br><br>▪ Lack of control of pricing models<br><br>▪ Slower adoption (the NSTIC's standard policies and procedures are still being developed) |

▪ **Social Security Administration (SSA).** The SSA allows taxpayers to establish an online account to view their estimated benefits, view earnings records, and apply for disability. Account registration requires that users provide an SSN, name, and address and answer a series of questions that allow for LOA-3 identity proofing (SSA, 2012). SSA's identity proofing, credential management, and per-session verification are supported by Experian and Symantec.

▪ **The Centers for Medicare & Medicaid Services (CMS).** CMS is in the process of developing an authentication system that will allow taxpayers to access their records, including accessing current and forthcoming state and federal health information exchange systems using a variety of devices, including mobile devices (Hickey, 2012). However, CMS is also planning to switch to third party credentials in the future (Walker, 2012). CMS's identity proofing, credential management, and per-session verification will be supported by Experian and Symantec.

Service providers, including government agencies, may decide to develop their own proprietary systems because of the ability to retain more control of security policies and procedures as well as the perception that the pricing set by IDPs may not be favorable. However, developing a proprietary system is far from simple. The difficulties can include the complexity and cost of developing an identity-proofing mechanism for citizens, the complexity and cost of providing citizens multiple services through a single authentication mechanism, and the difficulty in soliciting citizen adoption of credentials that will be accepted by only a single agency. Furthermore, the process of identity proofing individuals prior to conducting any transactions may be unnecessarily burdensome, as in USDA's case, which requires face-to-face interaction for credentials to be used with only one provider.

Alternately, the NSTIC vision offers improved authentication and reduced risk of identity theft at significant cost savings. For identity proofing, agencies can leverage individuals' credentials that have already been identity proofed (at zero or little cost, because the costs are shared across many accepting organizations). For credential management and authentication, agencies can rely on systems already established in the private sector, likely at a unit cost much lower than the cost of developing their own system. This estimated lower cost results from several factors:

- IDPs can spread their fixed and O&M costs for infrastructure across many RPs.

- As the market grows, third-party IDPs may gain expertise in identity proofing and credential management that would allow them to provide these services more effectively and efficiently than each government agency's security department and, therefore, may be able to develop and manage related activities at a lower cost and with higher quality.

- The customer service costs (e.g., when an individual loses his or her credentials) incurred by third-party IDPs are likely to be lower because economies of scale may allow third-party IDPs to provide customer service at a lower cost than a single organization.

- The ability to use the credentials to log in to a variety of government and business websites may result in fewer individuals losing such credentials.

Interviews with IDPs and identity experts suggest that confidence in the likely success of the NSTIC is high; however, successful implementation of an NSTIC-aligned authentication system that accepts third-party credentials is not guaranteed. Only two IDPs offering LOA-3 authentication currently have been approved (although several more are under review), and few citizens have these credentials as of yet.

NIH recently implemented the first system to accept FICAM approved LOA-2 third-party credentials to authenticate internet users of its web applications. NIH began accepting LOA-1 third-party credentials in 2010, and in 2012, it began to accept LOA-2 credentials through a pilot implementation with Verizon. NIH had accepted 120,000 unique externally issued credentials in 2011, and the number was expected to grow to over 400,000 in 2012. In interviews, NIH staff indicated that they were motivated by the desire to increase access to their digital assets, while at the same time not wanting to spend the time and money to develop their own authentication system that would require users to acquire a new credential.

In addition to the IRS's ability to develop its own proprietary authentication system and to accept third-party credentials, a third option exists. The Federal Cloud Credential Exchange

(FCCX) is a government initiative that aims to "provide all Federal agencies with an ability to accept the full range of FICAM-approved, non-federal government issued, credentials for online services for citizens" (U.S. Postal Service [USPS], 2012).[13] The FCCX allows federal agencies to use a single intermediary to verify the identities of their customers. The FCCX offers agencies additional cost savings over and above those likely to result from an IRS proprietary authentication system, because costs are distributed over numerous agencies and possible economies of scale are realized.

---

[13] The USPS released a Statement of Objectives as part of a 2012 Request for Proposals.

# 3. IMPROVED AUTHENTICATION AT THE IRS

The focus of this case study is a benefit-cost comparison of IRS acceptance of credentials issued by an IRS-only proprietary authentication system (Scenario 2) versus acceptance of NSTIC-aligned third-party credentials (Scenario 3). Currently, the IRS is developing a pilot implementation of a proprietary authentication system that would include identity proofing up to LOA-2 and ongoing management of credentials and authentication. The IRS is also considering accepting LOA-2 and LOA-3 third-party credentials in the future, although no timeline has been decided. In both cases, improving the security of online transactions would enable the IRS to increase its online service offerings and help it fight identity fraud. If taxpayers were able to use new IRS online services, the IRS could save money relative to providing the same services by phone, by paper, and in person. Taxpayers are expected to benefit from faster service.

In this chapter, drawing on insights from interviews with more than 30 IRS business units and publicly available reports, we describe how the IRS Online Services Group (OLS) plans to enhance its service offerings by improving online authentication and expanding online services.

## 3.1 IRS-to-Taxpayer Services: Current and Proposed

The IRS states as its mission to "provide America's taxpayers top quality service by helping them understand and meet their tax responsibilities and enforce the law with integrity and fairness to all" (IRS, 2011a). In addition to its tax collection duty, the agency is tasked with working with taxpayers to resolve any questions or issues regarding their taxes. This means that the IRS must devote a significant amount of resources to communicating and transacting with taxpayers. The discussion below focuses on tax returns submitted by and services provided to individual taxpayers and tax preparers submitting returns on behalf of individuals.

### 3.1.1 Current IRS Services

IRS services provided to individual taxpayers and paid preparers of individual income tax filings can be summarized as follows:[1]

1. **Authentication of taxpayers**

2. **Tax submission receipt and processing** by paper or in electronic format

3. **Addressing taxpayer general (nonaccount) inquiries** by mail, by phone, or in person

4. **Addressing taxpayer account inquiries** by mail, by phone, or in person

5. **Providing information and access to specific resources** via the website at www.irs.gov

---

[1] The list of activities was developed from interviews RTI held with the IRS.

6. **Addressing the grievances of taxpayers** experiencing economic difficulty, problems with their accounts, or complaints about IRS systems or procedures

7. **Providing the means for taxpayers to appeal to resolve tax disputes**

8. **Auditing and examining taxpayer accounts** for misreporting or identity theft

9. **Providing special services to victims of identity theft**, such as adding account indicators, requiring additional credentials, and conducting special screening of tax returns

10. **Initiating criminal investigations** of those who have not complied with tax laws and suspected perpetrators of financial crimes

These services are currently provided through paper-based communications, telephone communications, and online communications. In 2012, the IRS sent more than 250 million pieces of mail, and its staff made or received over 90 million phone calls (Table 3-1). Paper transactions often result in delays of several weeks or even months to complete because of mail delivery timelines and manual authentication processes.

The IRS services that could be affected by improved authentication can broadly be grouped into three categories: tax return processing, phone and mail communication with taxpayers, and identity theft prevention and mitigation.

*Tax Return Processing*

Based on the number of 2011 tax returns submitted as of July 30, 2012 (IRS, 2012e) and past tax submission trends, RTI estimates that approximately 80% of taxpayers will file returns electronically (or e-file) and 20% of taxpayers will file returns on paper for tax year 2012. Apart from those who choose to file on paper, approximately one million taxpayers (less than 1% of all taxpayers) are required to file their returns on paper because of the complexity of their returns. Paper tax return processing averages $3.41 more per tax return than electronic return processing. Furthermore, service times are much slower for paper versus e-filing; taxpayers who file electronically usually receive their tax refund within 3 to 4 days versus 3 to 4 weeks for paper filers.

**Table 3-1.    Summary of Estimated 2012 IRS Transactions with Taxpayers**

| Service | Type of Communication | | |
| --- | --- | --- | --- |
| | Paper | Telephone | Electronic |
| **Tax submissions** | 30 million paper tax submissions received | | 119 million e-filed tax submissions received |
| **Other services** | 220 million pieces of mail sent<br>Millions more received | 90 million phone calls made and received | Unknown |

Source: IRS interviews.

*Phone and Mail Communication with Taxpayers*

Today the vast majority of the IRS's communication with taxpayers occurs through the mail or by telephone. The IRS sends more than 220 million pieces of mail annually, including both notices (sent to update a taxpayer on an issue or issue resolution) and letters (sent for information purposes only), and receives millions of pieces of mail. The IRS also participates in more than 90 million phone calls per year based on requests made by taxpayers and calls made by IRS employees to resolve outstanding cases of tax fraud and tax money owed, for example.

*Identity Theft Prevention and Mitigation*

As discussed further in Section 3.2, identity theft costs the IRS and taxpayers a significant amount of time and money. The mitigation services provided by the IRS to taxpayers and additional prevention costs (models to identify fraud) and mitigation costs (criminal investigations) all aim to reduce the overall cost and impact of identity theft.

### 3.1.2 *Proposed IRS Service Changes*

In the future, the IRS plans to improve online services provided to taxpayers, with the goals of

- increasing the speed at which taxpayers receive individualized information and decreasing the cost of providing that information by enabling online taxpayer communications with IRS software applications and IRS staff (LOA-2 or LOA-3 will be required for all new online service applications);

- improving identity theft prevention and mitigation through improved online authentication for tax return submission (taxpayers will be able, but not required, to use new LOA-3 credentials for e-filing); and

- increasing the speed and reducing the cost of tax return processing by increasing interest in e-filing versus paper filing (LOA-3 will likely be required for e-filing with IRS proprietary credentials or third-party credentials).

New online service applications will enable information to be provided to taxpayers online, e-mails to be sent from the IRS to taxpayers online (and vice versa), and documents to be transferred from the IRS to taxpayers online (and vice versa). Planned applications that may require LOA-2 or LOA-3 are

- eTranscripts Phase II,

- Where's My Refund,

- Where's My Amended Return,

- ACH debit,

- secure webmail portal,

- manual notification tool,

- automated digital notification tool,

- live chat, and

- document transfer.

Improved authentication will have an impact on the identity theft activities of the IRS, including customer service activities. It is unknown what the specific impact will be, given that the use of LOA-3 credentials may not be required by the IRS for e-filing and paper returns will still be allowed. However, some cost reduction is likely, as is discussed further in Section 3.2.

E-filing also may increase as a result of improved online authentication. As noted above, processing electronic tax returns is much faster than processing paper returns. E-filing also results in taxpayers receiving their refund more quickly. The paper filing process consumes significantly more IRS labor for receiving, organizing, reviewing, transcribing, and filing paper returns. An increase in e-filing may occur when the IRS improves online authentication given that some taxpayers have cited security concerns as a primary reason for not e-filing. In the case of third-party credentials, additional e-filing adoption may occur as a result of the projected ease of use of e-filing with third-party credentials.

## 3.2 Current Authentication and Identity Management Activities at the IRS

To assess the impact of improved online authentication for the IRS, understanding current IRS authentication activities is essential. When working with individuals and paid preparers, the IRS must identify each party with whom they interact. Identification of all parties that file tax returns is done with a class of credentials referred to as a Taxpayer Identification Number (TIN). There are five types of TINs, but only three are relevant for the purposes of individual income taxes (IRS, 2012d):

- Social Security number (SSN)

- individual taxpayer identification number (ITIN)

- preparer taxpayer identification number (PTIN)

Figure 3-1 shows the identification and signing requirements for individuals and paid preparers for both paper and electronic filing. SSN, ITIN, and PTIN identifiers are used by the IRS as a primary method of identification both for tax return submission and all other communication, on paper and by telephone. The PTIN is issued by the IRS and used by tax preparers to access the IRS through the Registered Users Portal (RUP), a system in place to allow electronic tax submissions from individual tax preparers or tax preparation software programs.

**Figure 3-1.    Current Authentication within the IRS**

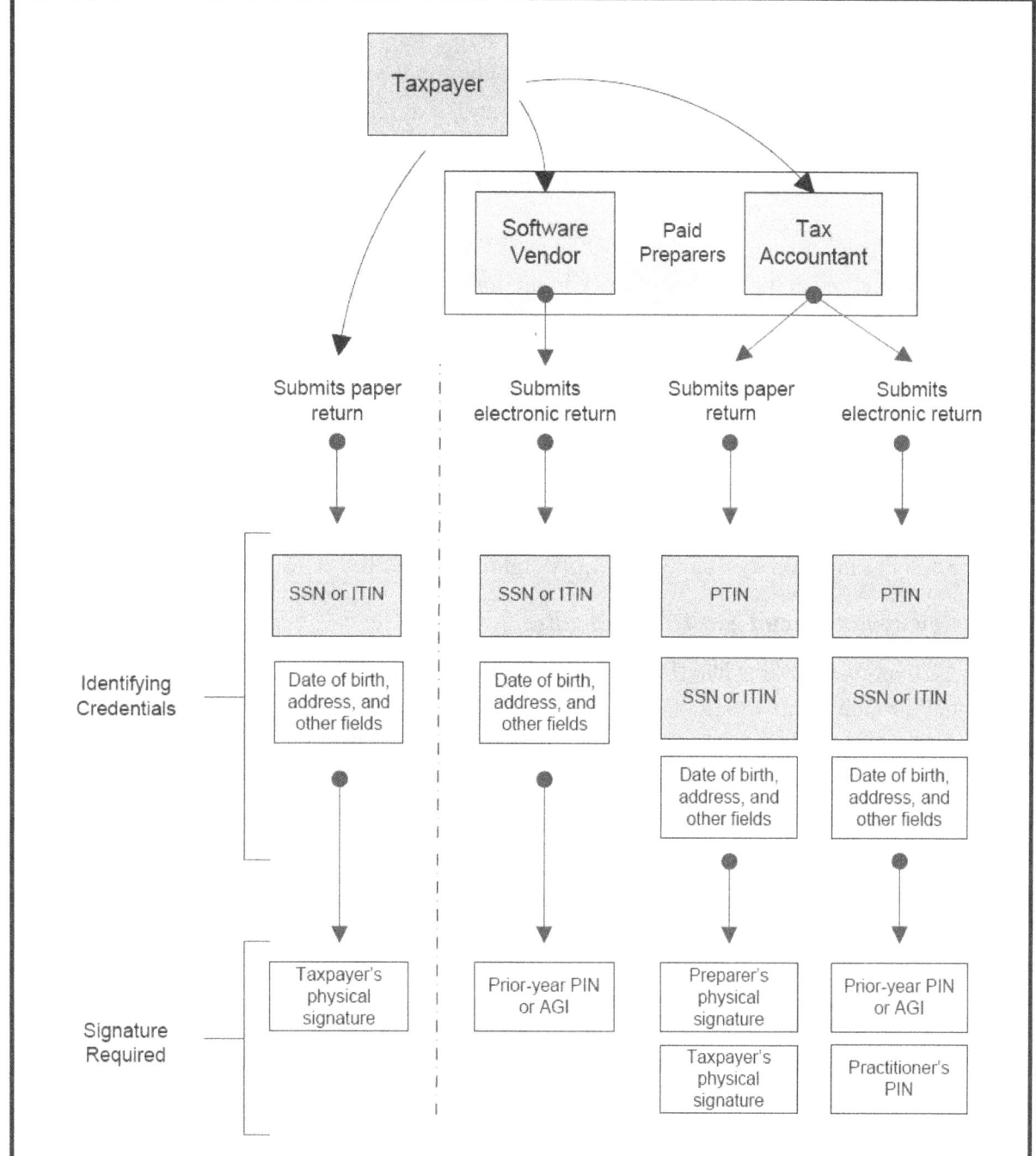

Note: The figure demonstrates the information a taxpayer and/or paid preparer must provide to authenticate a tax return. Within the "Identifying Credentials" bracket, the blocks in green represent primary means of identification.

An SSN is the most common credential used for taxpayer authentication, and an ITIN is assigned only to those who cannot get an SSN. The SSN is administered by the SSA, and the credential is used for identifying an individual to a variety of service providers (financial service and health care providers may ask for an SSN to verify identity). Because a citizen may use his or her SSN for a variety of activities, such as applying for credit or employment, one person's SSN is likely on file with several different service providers and written on countless paper forms.

Moreover, the IRS requires further proof of identity through the "signing" of a return. Paper returns are physically signed by the taxpayer and, if applicable, by a preparer; however, electronic returns have a different signing requirement. A returning filer can sign with a self-select PIN, a new e-file PIN created on irs.gov, or with last year's adjusted gross income (AGI).

Unfortunately, as explained below, the use of an SSN or ITIN contains a potential security flaw, and the extra layer of security in the form of an electronic PIN will not sufficiently address the inadequacy. Also, some IRS employees believe that the existing security measures do not encourage taxpayer confidence and limit the services that the IRS can provide online.

Finally, the authentication system involves sizeable costs. Both taxpayers and tax preparers sometimes lose or forget their credentials, and millions of taxpayers call an assistor to recover their PIN, username, password, or AGI as required for their tax submission. Generally, these calls are routed to an automated service, but on occasion, callers get a live assistor. In this case, the assistor must verify a caller's identity and retrieve a PIN.

### 3.2.1 Key Identity Fraud and Theft Activities

A direct result of this insufficient authentication is the level and scope of tax fraud committed through identity theft. The key credentials that the IRS relies on to verify the identity of a taxpayer can be stolen easily. According to a testimony to the House Committee of Oversight and Government Reform, taxpayers' SSNs and other details are generally stolen from information sources that are beyond the IRS's control, such as from the workplace, where the data are centrally stored (House Committee on Oversight and Government Reform, 2011). Although a PIN on an electronic return may provide a layer of security on top of the SSN, the House testimony showed that when an SSN is stolen, a PIN or AGI can be acquired by the thief as well. A thief can use the identifying credentials and file a paper return, bypassing the need for a PIN, and resulting in fraud.

If an identity thief acquires a taxpayer's credentials (such as SSN, name, address, date of birth, and income documents), two types of fraud can occur. In one scenario, the thief can file a return in the name of the legitimate taxpayer before the taxpayer files his own return. A thief will typically claim all possible credits to get the maximum possible refund, which will be sent to the thief or deposited into his account.

After the fraudulent refund is paid out, when the victim files his return, he is notified that he has already filed.[2] Upon verification of the victim's identity, he is awarded the appropriate refund. The refund given to the thief is a loss borne by the IRS. As discussed later, some of these losses are recovered, but many are not.[3] Adopting strong authentication solutions may make this type of identity theft and fraud more difficult online, especially if the taxpayer has the option to only allow tax filing with strong credentials. This type of identity fraud is shown in Figure 3-2.

In a second scenario, a thief may get a job, provide a stolen SSN to the employer, and ask that all or most taxes not be withheld. The taxpayer to whom the SSN belongs would not know about the other job in her name, and the income earned by the thief would fall under the responsibility of the taxpayer. As a result, the thief would not need to file and pay income taxes, because the tax liability would move to the victim. Because the victim is unaware of the additional income in her name, her return would be shown as underreported. After reporting identity theft and verifying their identity, the victim would be freed from the tax liability of the fraudulently reported income. In this case, because neither the thief nor the victim pays the full taxes due based on the misreported income, the IRS bears the loss. As this study focuses only on authentication for filing taxes, the solutions described herein are unlikely to have any direct impact on this type of identity theft or fraud. This type of identity theft and fraud is shown in Figure 3-3.

Although it is likely that neither the adoption of an NSTIC-aligned authentication system nor of a proprietary authentication system will directly affect the latter type of fraud, both systems are expected to have an indirect effect by enabling the IRS to concentrate its audit resources on a smaller subset of returns (those submitted without improved authentication credentials).

The biggest consequence of identity theft is the direct loss of money, discussed above, through fraudulently acquired refunds. In fact, according to a Congressional testimony by J. Russell George, Treasury Inspector General for Tax Administration, the IRS estimates that, each year, it issues $5.3 billion in fraudulent tax refunds (George, 2012). This number consists mostly of uncaught fraud identified by looking at patterns in tax returns ($5.2 billion) and also includes refunds caught after the fact when the victim filed a duplicate return ($70 million). In the latter case, the IRS is alerted to the fraud and is thus often able to recover many of the fraudulent tax refunds.

---

[2] Additionally, a particularly difficult type of fraud to identify occurs when someone files on behalf of a citizen who never files a return. For example, fraud is often committed by filing tax returns on behalf of children, criminals in jail, and retired persons. These individuals often do not file tax returns; thus, the fraud would never be identified.

[3] According to interviews, the IRS investigates identity theft rings based on suspicious patterns and outside leads, eventually recovering a fraction of the money lost to identity thieves.

**Figure 3-2.  Example of IRS Return Fraud: Filing Early**

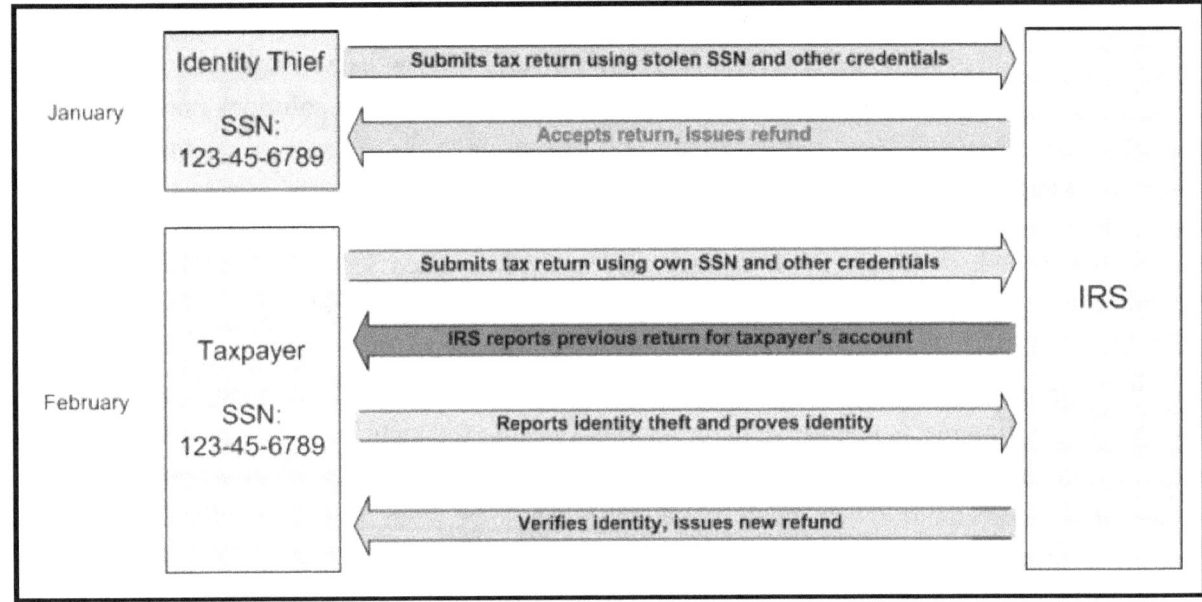

Note: This figure demonstrates return fraud, in which an identity thief files in January and receives a refund before the victim files in February. Once the victim files, he or she is informed of a duplicate return and must report identity theft and verify his identity before receiving the refund claimed. The solutions described in this study may help with this type of fraud online.

**Figure 3-3.  Example of IRS Income Fraud: Providing the Wrong SSN to Employer**

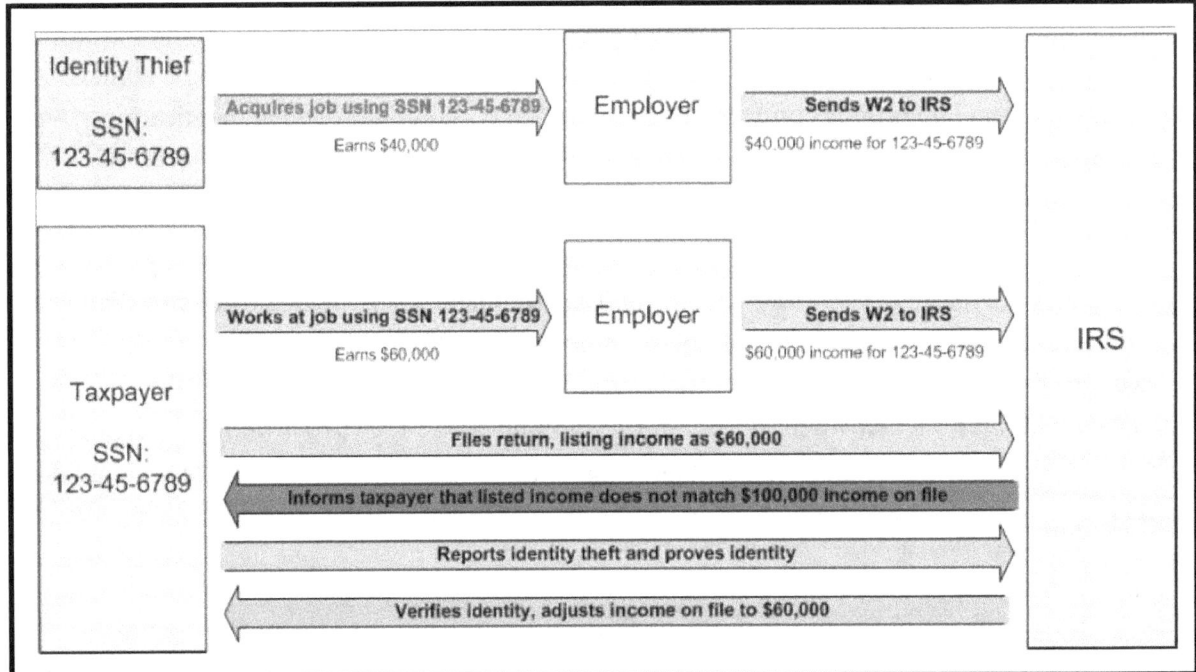

Note: The figure demonstrates income fraud. In this case, an identity thief earns income using the credentials of a legitimate taxpayer. This way, when the employers of both individuals send W2 forms to the IRS, the thief's income will be under the victim's name. When filing, the victim will unknowingly underreport her income and will need to prove to the IRS that the additional income on file with the IRS was reported unlawfully. The solutions described in this study are unlikely to directly reduce this type of fraud.

## 3.3 Options for Improving IRS Authentication

Currently the IRS is working to implement E-Authentication, which will support all future authentication activities. Between FY 2010 and FY 2012, the IRS has spent more than $20 million on internal labor, services, and software licenses to develop and operate the infrastructure needed to support their implementation of the E-Authentication framework, which will support improved authentication both of internal and external parties. Based on interviews with IRS staff we determined that some of this investment will support the development of infrastructure that could be used to accept third-party credentials, but much of this investment has been focused on developing the IRS's proprietary authentication solution.

Building on this infrastructure investment, the IRS has two main options to improve authentication:[4] (1) develop a proprietary online authentication system or (2) accept third-party credentials that have been provided by an IDP.[5] In both cases, the primary system costs will include

- identity proofing (LOA-2 and LOA-3),

- provisioning of robust multifactor credentials (LOA-3), and

- verification of each transaction (LOA-2 and LOA-3).

The difference is that under Scenario 2, the IRS would conduct each of these activities internally and not leverage the economies of scale offered by external organizations' user base and authentication systems.

Figure 3-4 is a conceptual model of electronic tax return authentication with either a new IRS proprietary credential or a third-party credential, as compared to a paper return. In the case of electronic returns with a new LOA-3 credential, the new credential would replace the need for the SSN, ITIN, or PTIN as the primary identifier. The new credential could also be used for non-filing services such as online services enabling account management.

Below we describe these two potential future scenarios: one in which the IRS manages all of these processes internally (including contracting with external service providers) through a new proprietary authentication system and one in which the IRS accepts NSTIC-aligned third-party credentials. We then compare these two scenarios and describe how several other government agencies are addressing their need for improved online authentication. In either case, online authentication will be much simpler.[6]

---

[4] Appendix B discusses several additional approaches identified during interviews that could improve authentication.

[5] The IRS could also decide to accept both types of credentials.

[6] Note that the IRS could decide to both manage its own proprietary authentication system and accept NSTIC-aligned third-party credentials.

**Figure 3-4.    IRS Authentication Using IRS Proprietary or Third-Party Credentials**

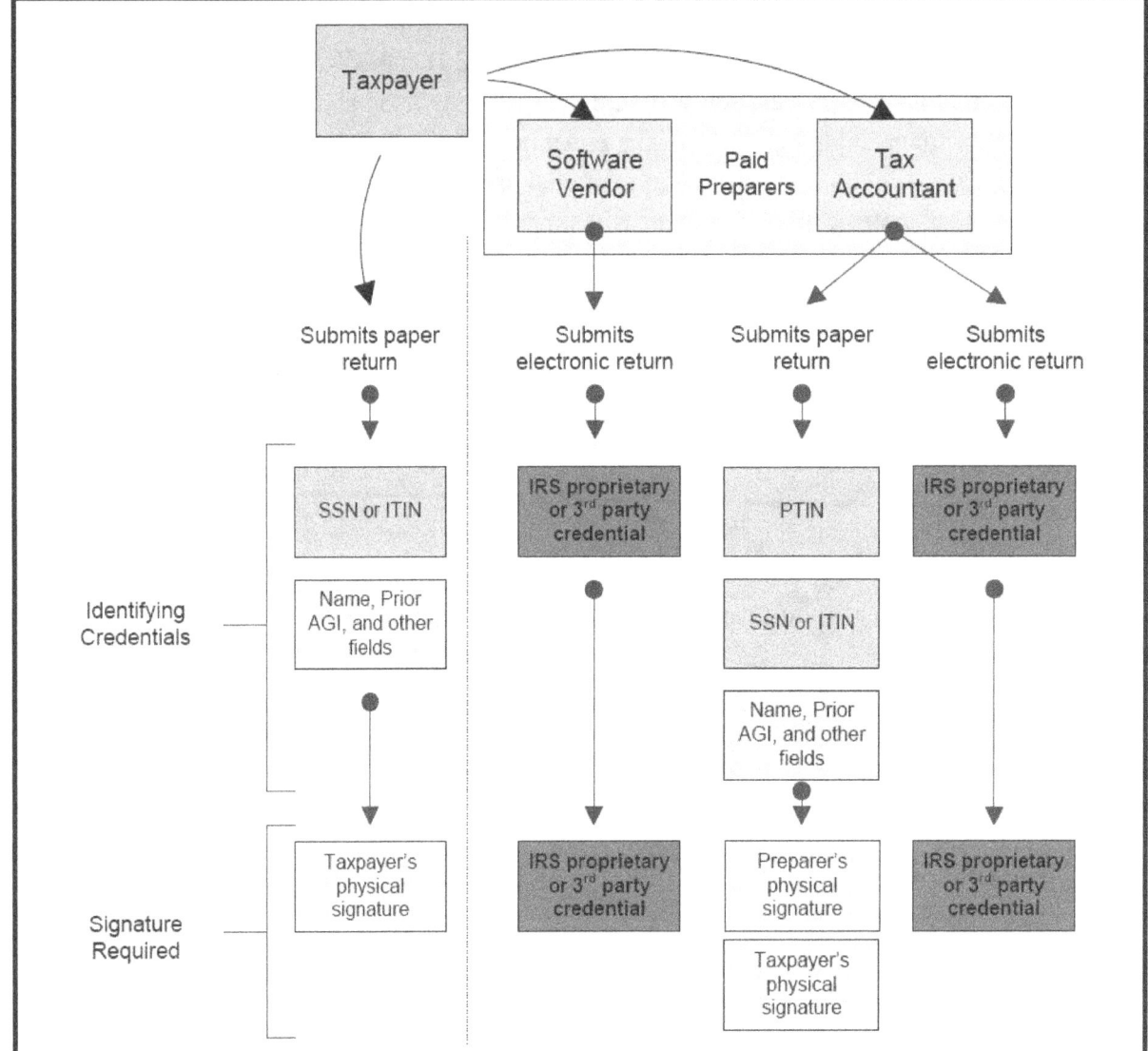

### 3.3.1  IRS Proprietary Authentication System

The IRS is currently implementing a pilot E-Authentication framework that could be responsible for centrally managing the proofing and enrollment of users, credentialing, identity management, and authentication. Broadly, the IRS's proprietary authentication system would perform the following functions:

- **Verify the identity of users** in a manner compliant with LOA-2 and eventually LOA-3.

- **Enroll users** into the system and create an identity for each user.

- **Bind each user's identity with a secure LOA-3 credential (including multifactor token)** and distribute each credential to the user.

- **Support users** who have forgotten their credentials and are unable to log in.

The IRS would need to verify the identity of all taxpayers who use their system, which may require combining IRS data with data provided by a trusted third-party organization to vet taxpayer identity confidently. The IRS plans to contract with one of the credit bureaus to supplement data the IRS has; combining these data sources, the IRS would be able to identity proof taxpayers.

Assuming verification succeeds, the IRS would need to store the user's information securely, and create an identity for that user. The identity would represent the user profile, containing any pertinent information that the IRS can use to identify and provide appropriate service to that user.

Once an identity has been proofed and established, the user would be assigned a credential, which under LOA-3 would be a combination of a username, password, and a cryptographic token. A cryptographic token can be stored in a smart card, mobile phone application, USB drive, or on a computer; the token can be distributed physically or transmitted electronically, depending on how it is stored. The IRS's current plan is to provide each user with a username and password and for a vendor to provide each user with a cryptographic token.

In addition to enrolling and credentialing users, the IRS would need to manage user accounts and support those users who have issues with their accounts. For instance, users may need to have their passwords reset or securely transmitted to them, and they may lose tokens, which would require a similar process of resetting and retransmission. The IRS currently plans for calls regarding access management to be routed to the IRS enterprise help desk.

Furthermore, two additional costs would be incurred. First, the IRS would need to connect new online applications to the new authentication system. The IRS's Online Services group would have to work with the E-Authentication group to develop applications that can be accessed only by properly authenticated users. Second, the IRS would need to manage any new threats that may arise. The IRS's Computer Security Incident Response Center (CSIRC), which is tasked with protecting IRS systems from cyber attacks, may be faced with new threats once a new authentication scheme is used, particularly if demand for new online services is high. If these new threats are considered to be a significant potential problem, CSIRC would have to update its prevention and mitigation methods accordingly.

### 3.3.2  *NSTIC-Aligned Authentication System*

For the IRS, making the decision to align with the NSTIC means that the IRS would have decided to accept taxpayers' third-party credentials, either to allow taxpayers to file tax returns or to interact with the IRS in some other way. For federal government agencies such as the IRS, NSTIC adoption requires compliance with FICAM, which primarily means that RPs (e.g., the IRS) can only accept third-party credentials from certified IDPs, and the IDPs must use federal government authentication standards, including interoperability standards. If the IRS decides to only accept third-party credentials,[7] the IRS will be outsourcing all of the functions listed above for the proprietary system to one or more vendors; however, they would still have the following activities and associated costs:

- Develop core infrastructure for the IRS to accept third-party credentials.

- Negotiate pricing and procedures with IDPs.

- Develop infrastructure for each application to connect to core authentication infrastructure.

- Anticipate and react to any new security challenges that may arise.

Although the level of infrastructure needed to accept third-party credentials would be much smaller than in the IRS proprietary authentication scenario, the IRS would need to build some infrastructure to accept third-party credentials. The IRS's E-Authentication group would be responsible for developing and managing the IRS's central authentication policies and procedures with regard to the NSTIC. One way of accomplishing this would be to build an "authentication gateway" through which a user will connect with a third-party credential before accessing online services.

Before third-party credentials can be accepted, the IRS would need to negotiate how fees are paid and how credentials from vendors will be transmitted and verified. The NSTIC NPO and the NSTIC Identity Ecosystem Steering Group are currently working to develop standard agreements aimed at making the RP and IDP negotiations very simple. We assume these costs are negligible and that the IRS would wait until standard agreements and pricing structures are in place before adopting the NSTIC.

As in the case of IRS-provided credentials, if the IRS decides to accept third-party credentials, they would need to develop the infrastructure to support each online application that would need to connect to the core authentication infrastructure. And the IRS would need to be prepared for potential security challenges and associated costs that may arise with the new system. For instance, there may be a possibility of individuals possessing false credentials mimicking those issued by third parties that the IRS accepts as genuine. Also, new channels of secure communication between the IRS and vendors providing authentication assertions could pose an increased risk as a new threat vector. If these new threats are considered to be a

---

[7] The IRS could decide to accept both proprietary IRS authentication credentials and third-party credentials, but for simplicity this discussion focuses on the case in which only third-party credentials are accepted.

significant potential problem, CSIRC would have to update its prevention and mitigation methods accordingly.

# 4. METHODOLOGY FOR BENEFIT-COST COMPARISON

Estimating the net benefits of the NSTIC to the IRS required employing benefit-cost analysis methods and identifying and justifying necessary assumptions.[1] In this chapter, we provide a broad description of the methodology used and assumptions made, delineate conceptually and empirically how the cost and benefit estimates were developed, and describe the data collection activities conducted.

## 4.1 Benefit-Cost Analysis Approach

Estimating the economic impact of a new technology or a novel technical approach, such as the NSTIC, requires taking a structured approach to understanding the potential costs and benefits and the likely level of user adoption. The costs and benefits of adoption should be calculated relative to the counterfactual, which for prospective studies is often assumed to be the status quo. If, for example, certain benefits are projected to materialize regardless of the new technology adoption, then the benefits should not be considered in the analysis.

In this case study analysis, the focus is on a single organization, the IRS, that is working on a pilot implementation of a proprietary authentication system and is beginning to discuss the potential for accepting NSTIC-aligned third-party credentials. As such, we defined the costs and benefits of two alternative future scenarios as compared to the status quo scenario (the base case) as follows:

- Scenario 1: Status Quo (IRS does not improve online authentication)

- Scenario 2: IRS Proprietary Authentication System (IRS issues and manages proprietary credentials and incurs proofing costs)[2]

- Scenario 3: NSTIC-Aligned Authentication System (IRS accepts third-party credentials but does not incur proofing costs)

Costs and benefits were first calculated for Scenarios 2 and 3 as compared with Scenario 1 (the base case) to access the net benefits. To estimate the specific NSTIC benefits, the economic impacts calculated in Scenario 3 were also compared against those calculated in Scenario 2.

Estimating the potential levels of adoption (demand) of IRS proprietary credentials and of NSTIC-aligned third-party credentials was not part of this study; however, it is a critical part of

---

[1] Note that this study did not seek to estimate the social benefits of the NSTIC. The focus here is on the benefits to the IRS. Additional benefits might flow to individuals—e.g., improved quality of service—and to other organizations—e.g., if the IRS's acceptance of third-party credentials results in increased adoption of other private and public organizations' online services authenticated through third-party credentials.

[2] Scenario 2 could also be the scenario in which the IRS accepts NSTIC-aligned third party credentials but the IDPs charge the IRS for the proofing costs.

the benefit-cost analysis. The potential costs and benefits in both Scenarios 2 and 3 will have an upper limit based of the total number of credential holders who might decide to use their credentials for a specific IRS online service. For Scenarios 2 and 3, three levels of demand were used to project a range of potential costs and benefits in a single year in the future: 20%, 50%, and 70% of taxpayers. Of those individuals who adopt new credentials, each specific IRS online service will have their own adoption curve (level of demand). Estimates of the adoption of each of the new IRS services were calculated based on information provided by the IRS.

The time frame in which certain types of costs and benefits will be incurred is a key component of accurately accounting for the accrual of costs and benefits of a new technology.[3] In this study, given the significant uncertainty about the timing of adoption, we provide estimates of the up-front adoption costs for each scenario, and then we provide estimates of the costs and benefits for a single year in the future for each of the three adoption ranges. All costs and benefits calculated are in 2012 dollars.

## 4.2    Taxonomy of Costs

To estimate the economic impact of Scenarios 2 and 3, relevant costs need to be identified and categorized. Broadly, Table 4-1 summarizes the major foreseeable adoption and ongoing costs required in Scenarios 2 and 3.[4] As shown in the table, most of the types of costs are the same for Scenarios 2 and 3, except that identity-proofing costs that will only be incurred in Scenario 2 and the cost to establish a relationship with third-party IDPs will only be incurred in Scenario 3.

### *4.2.1    Scenario 2: Up-Front Development and Implementation Costs and Ongoing Costs*

In this scenario, all of the cost categories listed in Table 4-1 are relevant.

*Up-Front Development and Implementation Costs*

First, the IRS will need to incur costs to develop the infrastructure needed to accept credentials issued by the IRS. Based on data provided by IRS staff, projected up-front costs for upgrading infrastructure are likely to be approximately $3 million.[5] This cost estimate is based on current and planned spending on system modifications. According to IRS staff, the investments being made would enable the IRS to accept either IRS proprietary credentials or third-party credentials with negligible modifications.

---

[3] When possible, all costs and benefits should be projected out several years (as far as is reasonable for a given technology and application environment), and the resulting stream of net benefits should be used to calculate metrics such as the net present value (NPV) of the technology and the benefit-to-cost ratio (BCR).

[4] These costs do not include up-front investments in infrastructure and process made by IDPs. We assumed these costs will be built in to identity proofing and per-transaction pricing models.

[5] Estimate provided by IRS Cyber Security staff.

**Table 4-1.    Primary Cost Categories, by Scenario**

| Type of Cost | Activity | Scenario 2 | Scenario 3 |
|---|---|:---:|:---:|
| Up-front fixed costs | ▪ Updating the core infrastructure to accept new authentication credentials | ● | ● |
| | ▪ Updating infrastructure to enable authenticated application use (per new application) | ● | ● |
| | ▪ Establishing a relationship with third-party IDPs | ● | ● |
| | ▪ Conducting identity proofing of each taxpayer | ● | ○ |
| Ongoing/ variable costs | ▪ Maintaining the core infrastructure and carrying out any changes necessary | ● | ● |
| | ▪ Maintaining the application-specific authentication infrastructure | ● | ● |
| | ▪ Providing customer support (e.g., reissuance of credentials) | ● | ● |
| | ▪ Conducting identity proofing of new taxpayers each year | ● | ○ |
| | ▪ Verifying credentials for each transaction session | ● | ● |

● = relevant cost category. ○ = not relevant cost category.

Once the IRS has built the necessary core authentication infrastructure, the IRS will need to integrate an initial set of applications—as well as future applications as they are deployed—with this new infrastructure so taxpayers can access each application using a third-party credential. In this case, an application refers to a specific online service considered to be separate from other online services.

Currently, the IRS plans for each new application to connect to a central authentication mechanism (the "authentication gateway"). Each application will need to be able to route the user through the authentication gateway. Interviews with IRS staff suggested the cost to integrate each new application with the authentication gateway varies widely, with estimates ranging from $163,000 to $1 million per application.[6] Data used for cost estimation for each application were based on application-specific data provided by the IRS or, when such data were not available, an average of the cost for similar applications.

Finally, in Scenario 2, the IRS will incur the cost of identity proofing all taxpayers who request new IRS authentication credentials. Estimates of the cost of identity proofing are based

---

[6] See detail data for each online service in Appendix C.

on the average of data provided during interviews with several IDPs. Based on interviews with IRS staff, we assumed that the costs will be similar for the IRS proprietary system.

*Ongoing Costs*

The IRS will incur costs associated with ongoing O&M of the core infrastructure needed to accept credentials or troubleshoot challenges. The IRS estimates the annual cost after year 1 to be approximately $500,000.

Each application will require an annual O&M cost. This cost consists of maintaining each application's interaction with the authentication gateway and other associated labor costs. The ongoing costs are approximately equal to 20% of the up-front costs for each application to be integrated with the authentication gateway.

Each year, the IRS will also incur the cost of identity proofing any new taxpayers who request new IRS authentication credentials. In tax year 2011, 6,328,945 individuals filed returns for the first time;[7] this figure was used for the analysis of future ongoing costs. Estimates of the cost of identity proofing are based on the average of estimates provided during interviews with several IDPs. The assumption is that the costs the IRS would incur are comparable to those that IDPs would incur.

The IRS will also incur costs associated with taxpayers using the IRS proprietary credentials. The costs for using the credentials will be a compilation of IRS labor and service and license agreements with vendors required to manage the authentication system.[8] Based on interviews with the IRS, we assumed these costs would be similar to the prices charged by IDPs for each transaction in which a third-party credential is used.

### 4.2.2 Scenario 3: Up-Front Adoption Costs and Ongoing Costs

In Scenario 3, third-party credentials are accepted by the IRS. In this scenario, identity proofing costs are included in the transactions costs charged by IDPs. The relevant up-front costs and ongoing costs are briefly described below.

*Up-Front Adoption Costs*

The up-front costs in Scenario 3 will be identical to those in Scenario 2, except that identity-proofing costs will not be incurred by the IRS. Here we assumed that the capability of a single credential to be used among multiple RPs motivates IDPs to absorb the cost of identity proofing individuals. Based on interviews with IDPs, they are likely to absorb the identity-proofing costs and recoup those costs through transaction fees. Half of the IDPs interviewed

---

[7] Data provided by the IRS in April 2013.

[8] If the IRS accepts third-party credentials in Scenario 2, a cost would be paid to external IDPs. Transactions costs could be packaged as an annual cost that the IRS pays per user (e.g., for unlimited transactions) or a fee charged to the IRS per transaction for each user-initiated online session.

indicated that 20% demand was a sufficient threshold for absorbing identity-proofing costs, while half said they may seek to pass along some identity-proofing costs until demand for third-party credentials approaches 50%.

In terms of infrastructure related spending, in this scenario, the IRS would need to develop a central authentication mechanism, as in Scenario 2; however, in this case, users seeking access to IRS online services would either be immediately redirected to the third-party IDP for credential verification or prompted for a credential to be routed to the IDP. Interviews suggest that the cost for this gateway will be similar to the costs in Scenario 2.

The up-front adoption costs for new infrastructure for each online service application are also assumed to be the same. In either case, each application will need to be able to route the user through the authentication gateway.

*Ongoing Costs*

The categories for ongoing costs will include the same core infrastructure and application-specific O&M costs as in Scenario 2, plus transaction costs for each transaction conducted with the IRS using NSTIC-aligned third-party credentials. Key differences are that the IRS bears no identify proofing costs in Scenario 3, and some ongoing costs are expected to be lower at higher demand levels.

## 4.3   Taxonomy of Benefits

The quantitative benefits to the IRS that would likely result from adopting new authentication credentials fall into three primary categories (see a brief description of each in Table 4-2):

- cost savings from new online services enabled by increased security (both Scenario 2 and Scenario 3)

- cost savings from reduced IRS authentication-related customer service activities (only Scenario 3)

- cost savings from lower average cost per transaction due to economies of scale realized by IDPs (only Scenario 3)

These benefits were quantified using data collected during interviews with IRS staff and through review of secondary data sources including IRS documents and websites. Although cost savings resulting from new, more efficient online services will accrue to the IRS in both Scenarios 2 and 3, cost savings from reduced authentication-related customer service activities will only accrue to the IRS in Scenario 3. If the IRS adopts a new proprietary authentication system, interviews suggest that the IRS will continue to incur the same authentication-related customer service activities as in the current system.

Although much less certain, additional benefits may result if NSTIC adoption results in an increase in e-filing (resulting in cost savings based on a reduction in more expensive paper tax return processing) or if NSTIC adoption helps to reduce identity fraud costs and losses. Table 4-2 describes these potential benefits, which were only addressed qualitatively in the study (see Section 5.4). The remainder of this section describes the benefit categories for which economic impact estimates were calculated.[9]

### 4.3.1 Cost Savings from New Online Services Enabled by Increased Security

The IRS has plans to deploy several new online services within the next few years. These services are at various stages of development. Table 4-3 lists and briefly describes the online services currently being proposed by the IRS that would require LOA-2 or LOA-3 credentials (see also Chapter 3) and the IRS's estimated level of adoption of these services.

**Table 4-2.    Overview of Benefit Categories**

| Area of Benefit | Benefit |
| --- | --- |
| *Benefits Quantified in this Study* | |
| Efficiency of online services | ▪ Reduction in call and mail volume for services that could be provided automatically online (Scenarios 2 and 3) <br> – Phone calls from taxpayers requesting account information and products (such as transcripts) to be mailed back <br> – Letters received from taxpayers for various account inquiries or requests <br> – Letters and notices related to account inquiries or special services (not return related) mailed to taxpayers |
| Reduced authentication-related customer service activities | ▪ Reduction in phone assistor support in dealing with (Scenario 3 only) <br> – Login issues to the RUP <br> – E-file PIN requests |
| Lower average cost per transaction under NSTIC | ▪ Lower average cost per transaction due to economies of scale realized by IDPs (Scenario 3 only) |
| *Benefits Discussed Qualitatively in this Study* | |
| Increased e-filing | ▪ Reduction in paper return processing costs <br> – Costs incurred prior to transcription into electronic format <br> – Costs of corresponding with taxpayers by mail as a result of errors |
| Identity theft reduction | ▪ Reduction in labor costs of mitigating cases of identity theft <br> – Cost of communicating by phone, by mail, and in person <br> – Criminal investigation of identity theft cases <br> ▪ Reduction in capital costs of phone calls and mail items |

---

[9] The benefits resulting from a lower average cost per transaction under an NSTIC-aligned system (Scenario 3) are included in the estimates of ongoing costs under each scenario and thus are not discussed further in this section.

**Table 4-3.     List of Planned New IRS Online Services and Estimated Level of Adoption**

| Online Service | Description | Estimated Future Level of Adoption, All Demand Levels |
|---|---|---|
| E-filing[a] | Allows taxpayers to submit their tax returns electronically using more secure credentials | 81%[b] |
| Online transcripts view-and-print capability (eTranscripts Phase II) | Allows taxpayers to request and receive their own tax return transcripts electronically rather than by mail | 38%[c] |
| Where's My Refund | Updates the existing "Where's My Refund" service, which notifies the taxpayer on the status of his or her refund, to provide more detailed information and offer return tracking | 38%[c] |
| Where's My Amended Return | Provides the ability for taxpayers to check the status of an amended return | 38%[c] |
| ACH debit | Allows taxpayers to pay taxes owed electronically rather than sending checks | 38%[c] |
| Secure webmail portal | Allows a taxpayer to manage an online account and communicate securely with the IRS | 40% |
| Manual notification tool | Allows for one-way electronic communication from IRS caseworkers to taxpayers with reminders and confirmation of account activity (e.g., receipt of documents) | 60% |
| Automated digital notification tool | Allows for the automated transmission of messages to the taxpayer based on account triggers | 50% |
| Live chat | Allow one customer service representative to have multiple conversations with individual taxpayers at once | 10% |
| Document transfer | Allows taxpayers to transfer documentation requested by the IRS online. | 30% |

[a] E-filing adoption is included because the IRS plans to allow new credentials to be used to e-file; however, additional related benefits were not quantified in this study.

[b] This projected level of adoption for e-filing is based on estimated tax-year 12 returns.

[c] These adoption estimates were not provided by the IRS. They were calculated by taking an average of the adoption estimates provided for all other online services.

The IRS was unable to identify how their adoption estimates would differ under each of the demand levels; therefore, the data in Table 4-3 were used for each demand-level analysis. For example, if there were 100 taxpayers, at the 50% demand level (50 taxpayers), 50% of these credentialed users (25 taxpayers) are expected to use the automated digital notification tool.

Currently, secure electronic communication between the taxpayer and the IRS is not available. Rather, all communication and account requests must be made by phone, by mail, or in person at a Taxpayer Assistance Center (TAC). Moreover, taxpayers needing to prove their identity must do so over the phone, by mail, or in person. Offline communication with the IRS includes

- labor costs for phone and TAC representatives interfacing with taxpayers;
- labor and administrative costs of inbound and outbound mail, including postage and processing/sorting labor;
- service costs of call volume (i.e., telephone service); and
- fixed costs, including overhead for call centers, TACs, and mail processing centers.

Each phone call or in-person visit requires that an IRS agent listen to the taxpayer and address his or her problem, and in some cases, the taxpayer may need to be redirected to other assistors. Also, mail correspondence involves similar steps but also requires labor to sort incoming mail and process outgoing mail. Moreover, corresponding with taxpayers involves administrative and services expenses including mail postage and telephone costs. These costs are expected to be proportional to the mail and call volume coming in and out of the IRS. Finally, fixed labor and capital costs are associated with managing all external communication centers within the IRS, and these costs will not change with significant fluctuations in offline communication.

*Benefit Estimation Assumptions*

A shift to online communication will lead to reductions in many current offline communication costs, including labor, administrative, and, in some cases, capital expenses. Two primary shifts are likely to occur: some information can be provided through self-service online applications without any IRS staff involvement, and many telephone and mail communications requiring IRS staff involvement can move to online communications.

Based on data provided by IRS staff, we assumed that successful implementation of online services will provide taxpayers a set of services, such as account management services and detailed account information that can be carried out without human interaction. If these online services are easy to use, taxpayers will be less likely to call the IRS or visit a TAC.

Moreover, secure online communication between the IRS and taxpayers may shift telephone and mail communications to online communications, reducing more expensive telephone and mail costs. Online communication could include secure e-mail and web chat. Although some labor will still be required, the total cost per interaction will go down significantly. As an example, assistors who operate on a web chat can run several chat sessions at the same time, leading to a more efficient use of labor.

*Estimation Procedure*

Our estimation of the net benefits of each new IRS online service differed based on the types of costs being replaced. However, for each online service, the following general formula was used to estimate the benefits:

$$\text{Online Service Benefits}_{ij} = (\text{Labor}_{tel} + \text{Labor}_{mail} + \text{Postage}) * \%\text{NSTICDemand}_j * \%\text{Adop}_{\text{Online Service } i}$$

where

| | | |
|---|---|---|
| $i$ | = | different online services, |
| $j$ | = | different NSTIC demand levels, |
| $\text{Labor}_{tel}$ | = | labor costs of telephone communication with taxpayers, |
| $\text{Labor}_{mail}$ | = | labor costs of mail communication with taxpayers, |
| Postage | = | cost to mail paper communication with taxpayers, |
| $\%\text{NSTICDemand}_j$ = | | estimated credentialed users at NSTIC demand level $j$, and |
| $\%\text{Adop}_{\text{Online Service } i}$ = | | the estimated level of adoption of online service $i$. |

## 4.3.2 Cost Savings from Reduced IRS Authentication-Related Customer Service Activities

With proprietary authentication solutions (the current system or if a new, more secure IRS proprietary authentication system is adopted), the IRS has to expend resources in supporting return preparers and taxpayers who have trouble with existing credentials and logging in. Currently, two main activities are associated with authentication support:

- supporting paid preparers (tax accountants) in IRS's RUP

- supporting taxpayers with an e-file PIN

Tax accountants and business taxpayers generally have special access to a set of IRS e-Services, which is "a suite of web-based products that will allow tax professionals and payers to conduct business with the IRS electronically," only available to "approved IRS business partners and not available to the general public" (IRS, 2012b). Accessed through the IRS's RUP, these services give return preparers and business partners special access to IRS services. To be granted this access, a user must register, a process that involves an identity-proving mechanism that mails a verification number to a specified address (does not meet LOA-2 requirements).

Registering to use the RUP and access IRS's e-Services is an online automated process, but users encounter problems and call the IRS for assistance. Moreover, users must manage a set of basic credentials (username and password), which are susceptible to loss. When a RUP user is unable to log in, he calls the IRS to help regain access to his account. The IRS then incurs labor and telephone costs.

Individual taxpayers e-filing on their own must also provide information for authentication purposes. Each electronically filed return must be signed with a PIN or AGI, and a taxpayer must remember her number from the previous year. When taxpayers lose or misplace their PIN or AGI, they often call to request a PIN replacement or to find out their previous year's AGI. This can be done over with phone with an automated system, with a live phone assistor, or over the web.[10]

### Benefit Estimation Assumptions

Accepting third-party credentials (Scenario 3) would enable the IRS to do away with many current authentication-related customer service costs—these costs would be borne by IDPs. Although IDPs would likely pass along such costs as part of the per-transaction fees charged to RPs such as the IRS, it is likely that these costs will be lower on a per-transaction basis; for example, if taxpayers are able to use their credentials at multiple websites, it is likely that they may lose or misplace their credentials less frequently than in the current IRS authentication system. If the IRS adopts a new proprietary authentication system (Scenario 2), the costs for customer service-related authentication activities would likely be higher than under Scenario 3.

According to data acquired from interviews with IRS staff, there are 219 assistors handling support requests from paid preparers, and out of this group, approximately 10 FTEs per year deal with authentication issues (e.g., account activation and login issues). With respect to e-file PIN requests, many of the requests are conducted online or with an automated system over the phone. From October 1, 2011, to April 7, 2012, there were 17.6 million completed requests either through an automated phone system or over the web. However, 225,406 calls were answered by a live assistor in that period. If the IRS adopts an NSTIC solution, we assumed that these calls will all be routed to the third party IDP, and that the associated costs for customer service will be included in the per-transaction costs charged to the IRS. Using overall call volume and labor cost data for e-services, we can develop an estimate of the labor cost for e-file PIN calls.

---

[10] Of note, according to interviews with IRS staff, the security of the process by which individual requests their PIN and AGI online is inadequate, likely resulting in identity theft that can be used to conduct identity fraud.

*Estimation Procedure*

Our estimation of the benefits of a reduction in current IRS authentication costs was calculated using the following formula for each year:

$$\text{Authentication Cost Savings} = \text{Labor}_{\text{auth}} * \%\text{Adop}_{\text{e-file}}$$

where

$\text{Labor}_{\text{auth}}$     =    labor costs of telephone communication related to authentication and

$\%\text{Adop}_{\text{e-file}}$   =    estimated level of e-filing with NSTIC-aligned third-party credentials.

## 4.4    Primary Data Collection Activities

We first conducted a series of scoping interviews with a variety of groups at the IRS to better understand the structure of the IRS and the categories of costs that it currently incurs, particularly those that could potentially be reduced through more secure online communication methods. These interviews helped us map out the types of processes that would be affected by more secure online communications, as described in Chapters 3 and 4 of this report.

Thereafter, we conducted more structured interviews with IRS employees, current and future IDPs, and other experts in online authorization and identity management. Through these interviews, we collected data to help estimate the costs of NSTIC adoption, the costs of the IRS proprietary system and the benefits of increased online security for the IRS. As part of these interviews, we spoke with staff at Oliver Wyman, who was contracted by the IRS to work on a study to identify potential solutions for reducing identity fraud within the IRS.

We interviewed IRS staff in the following groups/functions:

- Chief Financial Officer
- Compliance Data Warehouse
- Criminal Investigations Division
- Cyber Security
- Office of Compliance Analytics
- Office of Identity Protection
- Online Services
- RUP
- Statistics of Income Office
- Taxpayer Advocate Service
- Wage and Investment

From meetings with IRS staff, we received data on the number of transactions of various types within and outside of the IRS (e.g., the number of inbound and outbound calls and mail sent and received), as well as estimates of the costs associated with paper and electronic returns and taxpayer correspondence. These data on the number of transactions were used to estimate the annual per transactions costs and annual benefits in the analysis results presented in Chapter 5.

We have also used data from publicly available sources such as the most recent IRS Data Book (IRS, 2012c), reports by the Taxpayer Advocate Service, and Congressional testimonies. These sources provided us with precise, high-level volume and cost information regarding identity theft and limited data on offline taxpayer correspondence. However, parts of the data are ambiguous, and large gaps exist in the available data. The interviews helped discern the data that are relevant to this analysis.

# 5. ANALYSIS AND RESULTS

This chapter presents the results of the benefit-cost comparison of accepting proprietary credentials only for use with the IRS (Scenario 2) versus NSTIC-aligned third-party credentials (Scenario 3). The estimated costs and benefits presented in this chapter are presented as a one year "snapshot" of the future and are not attributable to any one point in time. The estimated costs and benefits are estimated with the assumption that the NSTIC vision has been realized and the IRS accepts NSTIC-aligned third-party credentials; that is, the hypothetical levels of adoption have been met. Throughout this chapter the data and data sources used to calculate the estimated costs and benefits are noted. Source data are also provided in Appendix C.

## 5.1 Scenario 2: IRS Proprietary Authentication System

As described in Chapter 4, three levels of adoption (demand) of IRS and third-party credentials were used to estimate costs and benefits: (1) 20% of the taxpayer base has an LOA-3 credential that can be used to log in to the IRS, (2) 50%, and (3) 70%.[1] Below, before reviewing the benefits and costs of the NSTIC-aligned solution, we describe our estimates for the IRS proprietary solution for each of the three demand levels.

### 5.1.1 Cost Estimates

In this scenario, the IRS accepts IRS-provided credentials for e-filing and online services and bears the full cost of the identity-proofing and transaction costs. For each adoption level, estimates of the per-taxpayer cost of identity proofing and the per-transaction cost were calculated based on interviews with IRS staff, other government agencies, security experts, and IDPs. Table 5-1 presents these demand levels and the corresponding costs to the IRS of identity proofing cost and transactions costs, as well as the estimate number of taxpayers and transactions at each level.[2]

Proofing costs are an up-front cost per taxpayer for the IRS (likely working with an outside vendor) to verify the identity of (or "identity proof") each taxpayer. These costs are assumed to be incurred in the process of proofing the identity of an individual (taxpayer) and are considered "one-time" costs. In contrast to these one-time costs, the IRS will incur "ongoing"

---

[1] Note that this study did not include an assessment of the likely levels of adoption in either Scenario 2 or 3. Instead, cost and benefit metrics were calculated for each of three hypothetical levels of adoption to provide a range of potential economic impact metrics.

[2] The estimated number of taxpayers at each demand level was calculated using the estimated number of taxpayers in 2012 as the total possible number of adopters. The number of transactions was calculated by assuming full adoption of e-filing and all online services for all adopters of new credentials—we assumed that use of each service would result in one transaction per year. Although this assumption likely overstates adoption for each online service, it likely understated the number of times each user of a specific online service logs in in to that service each year (likely more than once).

**Table 5-1.** **Average Identity Proofing and Transactions Cost, by Demand Levels**

| Cost Categories | Demand (% of taxpayers) | | |
|---|---|---|---|
| | 20% | 50% | 70% |
| Identity proofing cost (per person) | $1.37 | $1.22 | $1.07 |
| *Number of taxpayers* | *30,962,688* | *77,406,720* | *108,369,408* |
| Per-transaction cost (Scenario 2) | $0.07 | $0.05 | $0.05 |
| Per-transaction cost (Scenario 3) | $0.07 | $0.05 | $0.02 |
| *Estimated number of transactions* | *136,855,081* | *342,137,702* | *478,992,782* |

costs to verify the credentials of taxpayers each time they log in to IRS.gov to interact with the IRS using their proprietary credentials (e.g., to e-file or use one of the IRS's new online services). These costs could include IRS labor, software licenses, and/or service contracts depending on whether the IRS conducts the per-transaction verification themselves or whether they outsource the verification to an external vendor.[3]

Based on the average of data collected from identity providers, it is expected that as adoption of IRS credentials increases, the relevant average costs to the IRS decrease (Table 5-1). That is, as more taxpayers have credentials, the cost of identity proofing each person decreases and the per-transaction cost decreases across each demand level. However, based on average data collected from IRS interviews, in Scenario 2, the average per-transaction cost is expected to remain fixed at $0.05 per transaction after 50% adoption.

In addition to bearing the full cost of identity proofing for all credentials in this scenario, the IRS will also incur one-time adoption costs for the core authentication infrastructure and for each online service connected to the core infrastructure. Annual O&M costs will also be incurred for the core infrastructure and for each online service application requiring an IRS-issued credential, as well as annual identity proofing costs for new taxpayers who adopt credentials.

Under this scenario, total one-time adoption costs incurred by the IRS range from a low of $69 million under the 20% adoption level to a high of $139 million under the 70% adoption level (Table 5-2). Within the total one-time adoption costs, the costs to build the core authentication infrastructure ($3 million) and to build the infrastructure for nine online service applications ($25 million) are expected to contribute between 41% and 20% of total adoption costs, respectively. These two costs are expected to remain unchanged across all three demand levels.

---

[3] For simplicity the costs for credential verification are presented as per-transactions costs and annual estimates were calculated based on the projected number of transactions each year by demand level.

**Table 5-2.     Scenario 2: One-Time Adoption Costs, by Demand Levels ($000)**

| Adoption Costs | Demand (% of taxpayers) | | |
|---|---|---|---|
| | 20% | 50% | 70% |
| One-time identity proofing costs | $40,586 | $90,534 | $111,215 |
| One-time adoption costs (FICAM infrastructure) | $3,000 | $3,000 | $3,000 |
| One-time adoption costs (9 planned online service applications)[a] | $25,230 | $25,230 | $25,230 |
| **Total** | $68,816 | $118,764 | $139,445 |

Note: Sums may not add to totals because of independent rounding.  [a] See the breakout of adoption cost estimates for each of the 9 online service applications in Appendix C.

The largest contributor to total adoption costs is the identity proofing costs, which range from 59% ($41 million) to 80% ($111 million) of total adoption costs. As demand increases, one-time proofing costs contribute a larger share of total adoption costs.

Annual costs for each demand level comprise four ongoing costs: (1) O&M costs for the core infrastructure, (2) O&M costs for each online service application, (3) total annual transaction cost incurred for each time a taxpayer interacts with the IRS at IRS.gov, and (4) annual identity proofing costs for new taxpayers adopting credentials. For online services, the annual O&M costs (summarized in Table 5-3) are based on a series of "business case" data that the IRS provided for this study. The total annual costs range from $17 million (20% of taxpayers) to $25 million (50%) to $32 million (70%).

### 5.1.2   Benefit Estimates

The IRS is expected to realize significant cost savings from adopting new IRS LOA-2/LOA-3 credentials. Under Scenario 2, estimating annual benefits was limited to the cost savings borne from the increased use of online services and were calculated for the three demand levels to present a range of possible benefits. The following section presents the estimated cost savings from each demand scenario.

New online services are expected to generate significant cost savings for the IRS as labor and consumables (e.g., paper and postage expenses) expenses are reduced and taxpayer interaction at IRS.gov becomes more streamlined. Total estimated cost savings from increased use of online services range from a low of $91 million under the 20% demand level to a high of $318 million under the 70% demand level. Table 5-4 details the estimated cost savings for new online services

### 5.1.3   Total Up-Front Costs and Annual Net Benefits

Table 5-5 presents the estimated costs, benefits, and net benefits to the IRS from accepting new IRS credentials. Under Scenario 2, the up-front costs range from $69 million to

**Table 5-3.    Scenario 2: Annual Costs, by Demand Levels ($000)**

| Annual Costs | Demand (% of taxpayers) | | |
|---|---|---|---|
| | 20% | 50% | 70% |
| Annual O&M costs (core infrastructure) | $500 | $500 | $500 |
| Annual O&M costs (infrastructure for nine online service applications) | $5,046 | $5,046 | $5,046 |
| Annual transactions costs | $9,580 | $15,396 | $21,555 |
| Annual proofing costs | $1,730 | $3,859 | $4,740 |
| Total | $16,856 | $24,801 | $31,841 |

Note: Sums may not add to totals because of independent rounding.

**Table 5-4.    Scenarios 2 and 3: Estimated Annual Cost Savings from Increased Use of Online Services, by Demand Levels ($000)**

| Online Service | Demand (% of taxpayers) | | |
|---|---|---|---|
| | 20% | 50% | 70% |
| eTranscripts Phase II | $9,002 | $22,506 | $31,509 |
| ACH debit | $10,682 | $26,704 | $37,386 |
| Where's My Refund | $15,965 | $39,913 | $55,879 |
| Where's My Amended Return | $1,740 | $4,351 | $6,091 |
| Secure webmail portal | $30,910 | $77,274 | $108,184 |
| Manual notification tool | $1,000 | $2,500 | $3,500 |
| Automated digital notification tool | $17,154 | $42,885 | $60,039 |
| Live chat | $4,008 | $10,020 | $14,028 |
| Document transfer | $480 | $1,200 | $1,680 |
| Total | $90,942 | $227,354 | $318,295 |

Note: Sums may not add to totals because of independent rounding.

**Table 5-5.    Scenario 2: Annual Net Benefits, by Demand Levels ($000)**

| | Demand (% of taxpayers) | | |
|---|---|---|---|
| | 20% | 50% | 70% |
| Annual costs | $16,856 | $24,801 | $31,841 |
| Annual benefits | $90,942 | $227,354 | $318,295 |
| **Annual net benefits** | $74,086 | $202,553 | $286,454 |

Note: Sums may not add to totals because of independent rounding.

$139 million. Annual net benefits are the prospective cost savings less the expense incurred to achieve those cost savings. These net benefits range from $74 million per year under the 20% demand level to $286 million per year under then 70% demand level.

## 5.2    Scenario 3: NSTIC-Aligned Authentication System

Scenario 3 examines all of the cost-saving benefits resulting from the IRS's acceptance of LOA-3 credentials but, unlike Scenario 2, it does not include the one-time proofing costs for users. In addition, there are cost savings from reduced authentication-related customer service activities, and the ongoing transactions charges are expected to be lower at higher levels of demand.

### 5.2.1    Cost Estimates

In Scenario 3, the adoption costs are the same for the 20%, 50%, and 70% demand levels—$3 million in core infrastructure costs and $25 million in online services infrastructure costs. Table 5-6 details the estimated adoption cost under each of the three demand levels. Annual costs include $500,000 in O&M costs for core infrastructure; $5 million in O&M costs for online service applications infrastructure; and the annual cost of transactions—ranging from $10 million at 20% demand, $15 million at 50% demand, and $10 million at 70% demand.

**Table 5-6.    Scenario 3: Adoption Costs, by Demand Levels ($000)**

| Adoption Costs | Demand (% of taxpayers) | | |
|---|---|---|---|
| | 20% | 50% | 70% |
| One-time identity proofing costs | — | — | — |
| One-time adoption costs (core infrastructure) | $3,000 | $3,000 | $3,000 |
| One-time adoption costs (infrastructure for nine online service applications) | $25,230 | $25,230 | $25,230 |
| Total | $28,230 | $28,230 | $28,230 |

Note: Sums may not add to totals because of independent rounding.

**Table 5-7.    Scenario 3: Annual Costs, by Demand Levels ($000)**

| Annual Costs | Demand (% of taxpayers) | | |
|---|---|---|---|
| | 20% | 50% | 70% |
| Annual O&M costs (core infrastructure) | $500 | $500 | $500 |
| Annual O&M costs (infrastructure for nine online service applications) | $5,046 | $5,046 | $5,046 |
| Annual transactions costs | $9,580 | $15,396 | $9,580 |
| Total | $15,126 | $20,942 | $15,126 |

Note: Sums may not add to totals because of independent rounding.

### 5.2.2 Benefit Estimates

Under Scenario 3, benefit estimates include cost savings resulting from adopting new online services (as shown in Table 5-4 above), plus an additional category of benefits not applicable in Scenario 2—cost savings resulting from reducing authentication-related customer service activities.

*Cost Savings Associated with Reduced Authentication-Related Customer Service Activities*

Table 5-8 presents the expected benefits (cost savings) to the IRS from decreased authentication costs, including both

- authentication-related customer service calls (e.g., lost credentials) to support the RUP, which is used by tax preparers to submit tax returns, and

- authentication-related customer service calls to support the identity protection (IP) PINs (i.e., lost IP PINs), given to some taxpayers to increase security.

Total estimated benefits from decreased authentication costs range from approximately $0.5 million under the 20% demand level to almost $2 million under the 70% demand level. Decreased authentication costs resulting from a reduction in the number of live phone assistors dedicated to the IRS's RUP make up 24% of the total cost savings, while decreased authentication costs resulting from a reduction in the number of assistors dedicated to IP PIN requests make up the remaining 76% of total cost savings, under all three demand scenarios.

### 5.2.3 Total Up-Front Costs and Annual Net Benefits

In Scenario 3, the one-time up-front costs are approximately $28 million and the annual net benefits range from a low of $76 million under the 20% demand level to a high of $305 million under the 70% demand level. Estimated annual net benefits increase at all demand levels until reaching a maximum under the 70% demand level. Table 5-9 details the estimated annual net benefits accruing to the IRS under Scenario 3.

**Table 5-8.    Scenario 3: Estimated Annual Cost Savings from Decreased Authentication-Related Customer Service Costs, by Demand Levels ($000)**

| | Demand (% of taxpayers) | | |
|---|---|---|---|
| **Authentication Activity** | **20%** | **50%** | **70%** |
| Customer service calls related to the RUP | $136 | $339 | $475 |
| Customer service calls related to the IP PIN | $421 | $1,052 | $1,472 |
| **Total** | $556 | $1,391 | $1,947 |

Note: Sums may not add to totals because of independent rounding.

**Table 5-9.** **Scenario 3: Annual Net Benefits, by Demand Level ($000)**

| | NSTIC Adoption % | | |
| --- | --- | --- | --- |
| | **20%** | **50%** | **70%** |
| Annual costs | $15,126 | $20,942 | $15,126 |
| Annual benefits | $91,498 | $228,744 | $320,242 |
| **Annual net benefits** | $76,372 | $207,802 | $305,116 |

Note: Sums may not add to totals because of independent rounding.

## 5.3 Scenario Comparison

Two benefit-cost scenarios were developed to assess the impact of adoption of LOA-3 credentials within the IRS using three potential levels of demand for such credentials. Table 5-10 summarizes the costs and benefits in Scenarios 2 and 3 side by side to aid in comparing the results.

**Table 5-10.** **One-Time Costs and Annual Costs and Benefits, by Scenario and Level of Adoption ($000)**

| | NSTIC Demand | | | | | |
| --- | --- | --- | --- | --- | --- | --- |
| | 20% | | 50% | | 70% | |
| **Scenario** | **One-Time Costs** | **Annual Costs and Benefits** | **One-Time Costs** | **Annual Costs and Benefits** | **One-Time Costs** | **Annual Costs and Benefits** |
| **Scenario 1: Base Case** | $— | $— | $— | $— | $— | $— |
| **Scenario 2: IRS Proprietary Authentication System** | | | | | | |
| Benefits | | $90,942 | | $227,354 | | $318,295 |
| Total one-time costs | $68,816 | | $118,764 | | $139,445 | |
| *One-time adoption (infrastructure) costs* | *$28,230* | | *$28,230* | | *$28,230* | |
| *One-time identity-proofing costs* | *$40,586* | | *$90,534* | | *$111,215* | |
| Annual costs | | $16,856 | | $24,801 | | $31,841 |
| Annual net benefits | | $74,086 | | $202,553 | | $286,454 |
| **Scenario 3: NSTIC-Aligned Authentication System** | | | | | | |
| Benefits | | $91,498 | | $228,744 | | $320,242 |
| Total one-time costs | $28,230 | | $28,230 | | $28,230 | |
| *One-time adoption infrastructure) costs* | *$28,230* | | *$28,230* | | *$28,230* | |
| *One-time identity-proofing costs* | *$—* | | *$—* | | *$—* | |
| Annual costs | | $15,126 | | $20,942 | | $15,126 |
| Annual net benefits | | $76,372 | | $207,802 | | $305,116 |
| **Comparison of Scenario 3 with 2** | | | | | | |
| **One-time cost comparison (saved cost of Scenario 3)** | $40,586 | | $90,534 | | $111,215 | |
| **Annual net benefits comparison (additional benefits of Scenario 3)** | | $2,286 | | $5,249 | | $18,662 |

Note: Sums may not add to totals because of independent rounding.

The up-front adoption costs for Scenarios 2 and 3 differ significantly as a result of the fact that identity-proofing costs are not included in Scenario 3. Under Scenario 2, proofing costs account for 59% of estimated adoption costs under the 20% demand level, 76% of estimated adoption costs under the 50% demand level, and 80% of estimated adoption costs under the 70% demand level. The influence of proofing costs on estimated total cost is significant and grows with increased demand for more secure credentials. At 20% adoption, there is a $41 million difference in up-front costs, at 50% adoption it grows to $91 million, and at 70% adoption it reaches $111 million.

Annual cost savings range from $2 million to $19 million under Scenario 3.

## 5.4 Potential Additional Benefits to the IRS: Qualitative Discussion

Below two additional categories of benefits are discussed qualitatively because research conducted for this study concluded that there is significant uncertainty about the existence or level of several types of benefits that may come from the IRS's acceptance of NSTIC-aligned third-party credentials.

### 5.4.1 Reduced Costs and Losses Associated with Identity Theft

An electronic return that is protected by a strong third-party credential is less susceptible to identity theft than is a return filed using an SSN and signature or e-file PIN. As noted in Section 3.2.1, the IRS estimates that approximately $5.2 billion of fraudulently acquired refunds are issued each year. Of this estimate, approximately 85% of these fraudulently acquired refunds are requested through e-filing, while the remaining 15% are committed through paper return submissions.[4] Therefore, roughly $4.4 billion in identity theft-related tax fraud is estimated to be committed through e-filing and the remaining $800 million through paper return submissions.

Securing e-filed tax returns will help to combat the loss of IRS money to thieves claiming refunds they are not owed using identities that are not theirs. If third-party credentials become the de facto way in which tax returns are submitted electronically, the impact on IRS costs related to identity theft could be substantial. However, in the near-term, the IRS is not likely to make such a requirement of taxpayers.[5]

Illegitimate returns, defined here as those that have been submitted with stolen credentials, impose the following types of costs on the IRS:

- **Direct losses through duplicate returns and other fraudulent returns.** Direct losses from identity fraud constitute the biggest type of cost. This type of loss is

---

[4] These estimates of the percentage of identity theft conducted through e-filing are based on data provided by the IRS during interviews. These data are based on known cases of identity theft.

[5] Alternatively, if an individual could request filings and communications require stronger authentication, gains would be similar to those from the IRS mandating use of strong credentials.

represented by a refund paid out to a filer of an illegitimate return. The direct losses to the IRS include both losses associated with electronic returns (i.e., the total amount of fraudulent refunds given out to thieves who filed electronically) and losses associated with paper returns (i.e., the total amount of fraudulent refunds given out to thieves who filed on paper). This distinction is important, because the new third-party credentials would only make electronic returns more secure. Understanding the proportion of losses that stem from electronic returns will help us estimate the effects of the NSTIC on identity fraud.

- **Variable costs of labor associated with identity theft.** With each occurrence of identity theft, the IRS must take mitigating actions. These actions include assisting taxpayers who have had their identity stolen and investigating fraud cases. According to interviews with IRS staff, the agency is allocating approximately 717 employees to deal with identity theft. A large portion of these employees work to mitigate the effects of identity theft by fielding calls from taxpayers, verifying taxpayers' identities, and reissuing refunds. These employees handle identity theft directly and thus have an overall workload that directly relates to the identity-theft volume.

  More specifically, the following groups at the IRS would likely see staff reductions if the level of identity theft decreased: Criminal Investigations, Taxpayer Advocate Service, Identity Protection Specialized Unit, and Customer Service.

Because of the distribution of costs, any savings associated with reducing identity theft would be felt across many offices and business units. Direct losses could be expected to be proportionally reduced by the reduction in identity fraud. That is, a 20% reduction in identity theft could reasonably reduce 20% of losses due to payouts to identity thieves.

Variable labor spent reacting to cases of identity theft could be expected to be similarly reduced proportional to the reduction in fraudulent returns. For example, 10% fewer duplicate returns will eliminate 10% of notifications of a duplicate return, taxpayer calls reporting identity theft, correspondence to prove identity, and issuance of an IP PIN. With respect to criminal investigation, efforts are simply likely to switch to other, previously uninvestigated, cases.

Labor spent on proactive efforts to prevent cases of identity theft and other costs deemed to be fixed (largely unaffected by shifts in identity theft levels) were not included in the benefit estimates. Interviews suggest that these costs are typically strategic activities that likely do not correlate with specific amounts of anticipated identity theft. That is, the establishment of a new strategy, initiative, system, model, or filter to prevent fraudulent returns is at the discretion of the responsible business unit, and the decision would be largely independent from the anticipated number of fraudulent returns going through the IRS. Thus, these costs are not included in our analysis.[6]

The reduction in costs associated with a reduction in identity theft depends on the specific changes in identity theft as a result of more secure online transactions. Because third-party credentials will be used only for electronic returns, the NSTIC solution does not protect paper filing. There are two ramifications of this:

- Identity theft through electronic returns may decline in the near term because legitimate e-filers are protected by a third-party credential; however, any significant reduction in fraud would require that taxpayers can opt-in to require stronger authentication when e-filing to prevent fraud on their account.

- Identity theft associated with paper returns may increase because it will be easier to commit fraud on paper returns compared with electronic returns.

Third-party credentials will be used only by taxpayers who file electronically. This means that paper filers, whose only layer of security is a physical signature, will not see the added security benefit, and the IRS may not see an overall reduction in identity theft-related fraud committed through paper return submissions. However, the IRS may be able to target their resources better because they will not need to investigate returns submitted via LOA-3 credentials for potential identity fraud.

Given that 85.7% of IRS identity fraud is estimated to occur electronically,[7] identity fraud losses may decrease briefly in the short run but may not in the long run within additional controls implemented by the IRS. In the long run, it is likely that some skilled or persistent thieves will switch to paper methods of committing tax fraud through identity theft, while the less-dedicated identity thieves will give up. Understanding the types of fraud will allow the IRS to estimate how much of fraud shifts and how much of it disappears. This analysis assumes that no shift will occur because there is no reasonable way to estimate this shift.

### 5.4.2  Reduction in Paper Return Filings

Because of the increase in security and improved ease of use of new credentials, individual taxpayers that have traditionally filed paper returns may have greater incentive to file

---

[6] The fixed costs in preventing identity theft are a smaller part of the overall labor costs compared with variable labor costs. Through interviews, we have identified 25 employees in the Office of Identity Protection, which is responsible for agency initiatives to prevent ID theft and overall identity theft strategy. This labor effort is assumed to remain the same regardless of the volume of identity theft. In addition to this number, approximately 216 employees are working with Criminal Investigation, whose job it is to investigate cases of identity fraud (Miller, 2012). This is considered a fixed cost because investigative labor's workload is limited by its resources, so not all identity theft cases can be investigated. We assume that if overall identity theft decreases, then Criminal Investigation will focus on other, previously lower priority, cases of identity theft.

[7] Based on data on known cases of identity theft provided to RTI by the IRS.

electronically (e-file).[8] Taxpayers who once filed on paper because of concerns about security may now shift to electronic filing (IRS, 2008). If more taxpayers do file electronically, the IRS will save money as paper returns, which are relatively costly to process, are gradually replaced by electronic returns that bypass a majority of the steps in processing returns. Conversely, some users may perceive negatively additional security steps or changes to current authentication schemes, decreasing the propensity to e-file. Because of these effects, future observed changes in taxpayer behavior because of changes in authentication schemes may differ from the estimates used in this analysis.

In tax year 2012, approximately 80% of tax returns were submitted through e-filing. According to a 2008 study commissioned by the IRS to assess how to increase the level of e-filing, the remaining 20% could be broken out into 4% who are "late adopters"—individuals who are likely to shift to e-filing sometime in the future but would do so more quickly with extra incentives—and 16% who are "laggards"—individuals who will likely not adopt e-filing unless it is the only option available (IRS, 2008). Through surveys and interviews, this study found that many individuals who were not e-filing were concerned about the security of doing so and perceived e-filing as difficult to use. If the IRS is able to convince some of the "late adopters"— possibly as many as 5.8 million taxpayers—that NSTIC-aligned third-party credentials or new IRS proprietary credentials will increase the security and possibly increase the ease of use of e-filing, an increase in e-filing may result, further increasing the benefits of the NSTIC or IRS proprietary authentication.

The clear cost differences between e-filing and paper filing are easily seen by reviewing the IRS's submission processing pipeline (see list of key steps in Figure 5-1). This process begins with the delivery of paper and electronic returns and ends with the posting of the return to the IRS Master File. Whereas paper returns need to be processed, electronic returns arrive fully processed and formatted. Therefore, electronic returns skip a portion of the pipeline (Steps 1–5 in Table 5-1). After a paper return is transcribed into electronic format, the pipeline is identical for both types of returns. As a result, the following costs can be eliminated for each return filed electronically instead of on paper:

- reductions in labor for sorting, batching, and stamping returns

- reductions in labor for examining paper returns, with the assumption that there are employees who check only paper returns

- reductions in labor for transcribing returns onto electronic media

---

[8] From a taxpayer's perspective, e-filing with a new proprietary credential or NSTIC-aligned credential may be more time consuming the first time if the sign-up process (including identity proofing) is required. However, assuming the credentials can be used for more services at the IRS and/or elsewhere (in the case of NSTIC-aligned credentials), taxpayers may find it easier to keep up with their new credentials as opposed to finding their PIN, username, password, or AGI.

**Figure 5-1.** **IRS Tax Return Submission Processing Pipeline: Paper Return Submission Steps Eliminated by Electronic Return Submission**

Paper Tax Return Steps Made Unnecessary by e-filing

**1. Delivery, Extraction, Sorting**: Incoming returns in barcoded envelopes are sorted, opened, and separated by machines.

**2. Batching**: Returns are further sorted by employees, grouped into batches, which are marked and logged into a computer.

**3. Coding and Edit**: Tax examiners look for errors on the paper returns and assign codes to facilitate data entry. Examiners may also correspond with taxpayers for corrections. Examiners typically check for obvious errors such as missing signatures, missing documents, or misplace entries.

**4. Sequential Numbering**: Batches are stamped and numbered.

**5. Transcription**: Paper return data are transcribed into a computer, and parts are re-transcribed by another employee to verify transcription.

**6. Error Resolution System**: Electronic return data are checked for errors. Returns that are not correctable by examiners are sent back to the taxpayers for correction.

**7. Posting**: If no errors or inconsistencies are found and all requirements are met in the return, the return is posted to the IRS Master File, where the return may be further scrutinized by other groups or a refund would be posted.

Source: The steps are outlined in a video published by the IRS and available here: http://www.irsvideos.gov/Professional/IRSWorkProcesses/SubmissionProcessingPipeline.

- reductions in ongoing maintenance costs of machines that sort, open, and separate paper returns

Cost data are available from the IRS's *Cost Estimate Reference FY 2011* (2012a), which estimates that the average cost of processing a single paper return is $3.55. In contrast, the average cost of processing a single electronic return is $0.14. The cost breakdown is shown in Table 5-11. This table provides a short-run representation of the costs of processing paper versus electronic returns and is directly affected by any potential reduction in paper filing.

That is, if one million taxpayers were to switch from paper to electronic filing, the IRS would save $3,550,000 in paper return processing but would incur a cost of $140,000 in electronic return processing, a savings of $3,410,000. These cost savings would be manifested in

**Table 5-11.** **Average Per-Return Cost for a Paper 1040 and Electronic 1040, by Cost Category**

| Cost Category | Cost per Paper 1040 Return | Cost per Electronic 1040 Return |
| --- | --- | --- |
| **Total cost** | $3.551 | $0.143 |
| **Pipeline processing activities** | $1.308 | $0.039 |
| Transcription | $0.500 | $0.005 |
| Error resolution system | $0.325 | $0.021 |
| Coding and edit | $0.255 | — |
| Delivery, extraction, sorting, batching, and numbering | $0.196 | — |
| Archiving | $0.030 | — |
| Accounting | $0.002 | — |
| Electronic form processing | — | $0.014 |
| **Nonpipeline processing activities** | $0.097 | $0.020 |
| Notices to taxpayers | $0.064 | $0.009 |
| Entity | $0.021 | — |
| Resolving unpostable returns | $0.012 | $0.011 |
| **Other labor costs** | $2.147 | $0.083 |
| Overhead | $0.837 | $0.032 |
| Benefits | $0.776 | $0.031 |
| Management and supervision | $0.489 | $0.018 |
| Quality assurance | $0.043 | $0.002 |

Source: IRS (2012a).

the reduction of return-processing staff, which could then be redistributed to other IRS functions.[9]

---

[9] Moreover, long-run changes are associated with a reduction in paper filing. During interviews with IRS staff, we learned that the IRS has shut down several paper processing centers because the reduction in paper filing reduced their utilization. Thus, with a stable drop in paper returns, cost savings may occur through shutting down machinery or entire processing centers. However, these costs are not in the scope of our study, because they are long-term strategic decisions rather than direct consequences of a shift to e-filing.

# 6. CONCLUSION

The NSTIC lays out a vision of a future in which online authentication can be conducted efficiently and effectively, resulting in reduced operational and customer service costs to organizations, improved services for individuals, and reduced security-related costs to both organizations and individuals. In this system, users could have their identity verified by one IDP and then use the credentials they receive to interact with many organizations and websites, decreasing the need for new credentials for every interaction.

This study conducted an analysis of the costs and benefits to the IRS of two FICAM-compliant authentication alternatives. Costs and benefits were estimated for three scenarios: (1) IRS maintains the status quo, (2) IRS accepts IRS proprietary credentials, increases the number of services offered online and the pays the cost of identity proofing all users, and (3) IRS accepts NSTIC-aligned third-party credentials and increases the number of services offered online but does not pay the cost of identity proofing users. This three-scenario structure allowed several comparisons:

1. **Scenario 2 (IRS Proprietary) versus Scenario 1 (Status Quo)**: As compared with the status quo, if the IRS launches a proprietary system, it will cost approximately $69 million to $139 million up front, but the annual net benefits at adoption levels of 20%, 50%, and 70 % will be $74 million, $203 million, and $286 million respectively.

2. **Scenario 3 (the NSTIC) versus Scenario 1 (Status Quo)**: As compared with the status quo, Scenario 3 will cost approximately $28 million up front, but the annual net benefits at adoption levels of 20%, 50%, and 70 % will be $76 million, $208 million, and $305 million, respectively.

3. **Scenario 3 (the NSTIC) versus Scenario 2 (IRS Proprietary)**: Comparing two future alternative scenarios, Scenario 2 will cost between $41 million and $111 million more up-front than Scenario 3 as a result of the cost of identity proofing users. The annual net benefits will be between $2 million and $19 million higher under Scenario 3, as a result of the benefits accruing from decreased authentication-related customer service costs in Scenario 3 and the higher per-transaction costs at the 70% adoption level in Scenario 2.

Scenario 3, in which the IRS accepts third-party credentials and does not pay identity proofing costs, is the ideal vision of the NSTIC. This scenario will result if taxpayers acquire LOA-2 or LOA-3 (depending on the LOA required by the particular application) third-party credentials prior to interacting with the IRS or if IDPs do not charge the IRS for identity proofing. In contrast, in Scenario 2, the IRS decides to accept credentials that are proprietary—generated and managed by the IRS—and as such the IRS pays identity-proofing costs.

Based strictly on the cost and benefit metrics, the IRS should accept NSTIC-aligned, third-party credentials if they do not have to pay for identity proofing regardless of the levels of demand. It is important to note that in Scenario 2, if the IRS decides to generate and manage its own credentials, the level of adoption of such credentials may not be as high as if the IRS decides to accept third-party credentials, assuming the IRS does not require proprietary credentials.[1] Third-party credentials can be beneficial to users—such credentials can be used to access any RP website that accepts them, so users who acquire third-party credentials may not have to acquire new credentials to access multiple websites, thus saving them time and reducing the risk of data being stored in multiple locations. Given these potential benefits to users, third-party credentials may have higher levels of adoption than proprietary credentials. The IRS should take this into consideration when deciding whether to launch their own proprietary authentication system versus accepting third-party credentials.[2]

The benefits quantified primarily stem from new online services that the IRS plans to launch. Cost savings related to more efficient authentication are not projected to result in sufficient benefits to outweigh the costs of adoption. However, not all the potential benefits of the IRS's adoption of the NSTIC were quantified. The following potential benefits to the IRS could result from the NSTIC adoption:

- cost savings if identity fraud costs or losses could be reduced

- cost savings if e-filing increases as a result of the NSTIC

Furthermore, the benefits to taxpayers of NSTIC-aligned third-party credentials were not calculated. These benefits could include the following:

- cost savings from reduced identity theft and fraud

- time savings from reduced identity theft and fraud

- time savings from faster services provided online

- utility (taxpayers' perceived benefit) of additional information available online

In 2011, 641,052 taxpayers were identified by the IRS as being victims of identity theft, and the total number of victims is likely higher. Most of these taxpayers eventually received the correct refund, as appropriate; however, the time investment and waiting time required for resolution can be significant, in many cases longer than one year (Treasury Inspector General for Tax

---

[1] The IRS could decide to require proprietary credentials or NSTIC-aligned credentials, but our conversations with IRS staff suggest that these are not likely policies in the near term because any such change may result in a decrease in the number of taxpayers e-filing.

[2] As part of the FCCX, the government also is setting up a system to allow federal agencies to use a single intermediary to verify the identities of their customers. As such, the FCCX may offer agencies additional cost savings over and above those likely to result from an IRS proprietary authentication system, because costs are distributed over numerous agencies and possible economies of scale are realized, while still allowing RPs more control than relying solely on third-party credentials.

Administration, 2012). The monetized time savings associated with reduced identity theft could significantly increase the estimated benefits of NSTIC-aligned credentials.

## 6.1 Role of Technical Infrastructure

The need for technical infrastructure to support improved authentication is significant. The lack of sufficient authentication causes a market failure in online transactions as a result of the *excessive transactions costs*. These excessive transactions costs generally include an opportunity cost, increased cost of products (as a result of authentication costs), increased financial losses via theft or fraud (costs imposed on individuals and organizations), and increased customer service costs.

In the case of third-party authentication, several types of market failure are at play. A *coordination market failure* exists given the significant coordination needed to develop standards, protocols, and standard operating procedures that multiple stakeholders agree on so that individuals and organizations that use the authentication system can do so with relative ease based on system efficiency and widespread interoperability. A second coordination failure results in the form of a *chicken and egg problem* because multiple RPs need to build systems to accept third-party credentials to incentivize potential users to acquire third-party credentials and incentivize IDPs to develop identity-proofing and identity management systems.

The private and public sectors have a role in trying to address this market failure, which manifests both as a technical and economic barrier to adoption of third-party credentials. NIST has a central role to play in this process.

## 6.2 Recommendations for the NSTIC NPO and NIST

Understanding the economic impact of the NSTIC is complicated and a one-size-fits-all case study is not possible. However, this case study identified that increased efficiency and associated cost savings from moving customer service functions online can easily justify several potential NSTIC adoption scenarios. The impact of the identity-proofing and per-transaction costs charged by third-party credential providers (e.g., IDPs) could have a significant impact on the net benefits calculations. Additional investigation to discern likely private-sector pricing models and public-sector IDP pricing models could greatly improve forecasting abilities of potential adopters.

Understanding why and how individuals will decide to adopt NSTIC-aligned third-party credentials is also critical. This study made assumptions about adoption based on the best available data, but to overcome the chicken and egg problem, studying methods to solicit individual adopters must be included in the NSTIC planning efforts.

Finally, although this study did not aim to catalogue the specific technology infrastructure needs to support the NSTIC, interviews with IRS staff and external identity experts and additional research suggest that a variety of needs still exist, mainly for the private sector:

- standard agreements/standard operating procedures concerning stakeholder relationships (e.g., IDPs and RPs)

- standards, protocols, and policies dictating how and what information will be exchanged between stakeholders

The Identity Ecosystem Steering Group called for in the NSTIC is working on these and other issues, but additional investment by NIST or others still may be necessary to address the remaining technology infrastructure needs to ensure the NSTIC's success.

Improved authentication could significantly reduce transaction costs for online communications; however, both private and public investments are needed to help reduce the costs of authentication. NIST, industry associations, and consortia must continue their work coordinating the relevant stakeholders, facilitating information sharing, and providing objective subject matter expertise.

# 7. REFERENCES

112th Congress. (2011–2012). (n.d.). *House report 112-284*. Retrieved from http://thomas.loc.gov/cgi-bin/cpquery/?&dbname=cp112&sid=cp1128FKTw&refer= &r_n=hr284.112&item=&&&sel=TOC_622405&

Blue Research. (2011). Consumer perceptions of online registration and social sign-in: US consumer market research. Prepared for Janrain.

Bolten, J. B. (2003, December 16). E-Authentication guidance for federal agencies. *Executive Office of the President*. Retrieved from http://www.whitehouse.gov/sites/default/files/omb/memoranda/fy04/m04-04.pdf

Burr, W. E., Dodson, D. F., Newton, E. M., Perlner, R. A., Polk, W. T., Gupta, S., & Emad, A. N. (2011, December). *Electronic authentication guideline*. National Institute of Standards and Technology. Retrieved from http://csrc.nist.gov/publications/nistpubs/800-63-1/SP-800-63-1.pdf.

Burr, W. E., Dodson, D. F., & Polk, W. T. (2006, April). *Electronic authentication guideline*. National Institute of Standards and Technology. Retrieved from http://csrc.nist.gov/publications/nistpubs/800-63/SP800-63V1_0_2.pdf

Federal Chief Information Officer Council. (2011, December 2). *Federal identity, credential, and access management (FICAM) roadmap and implementation guidance*. Retrieved from http://www.idmanagement.gov/documents/ FICAM_Roadmap_and_Implementation_Guidance_v2%200_20111202.pdf

Florencio, D., & Herley, C. (2007). *A large-scale study of web password habits*. World Wide Web (WWW) Conference 2007, May 8–12, Banff, Alberta, Canada. Retrieved from http://research.microsoft.com/pubs/74164/www2007.pdf

George, J. R. (2012). *Identity theft and tax fraud*. Joint Hearing before the Committee on Ways and Means Subcommittees on Oversight and Social Security. Retrieved from http://waysandmeans.house.gov/UploadedFiles/George_Testimony.pdf

Hickey, K. (2012). *Medicare, Medicaid implement online anti-fraud technology*. AOL Government. Retrieved from http://gov.aol.com/2012/03/15/medicare-medicaid-implement-online-anti-fraud-technology/

House Committee on Oversight and Government Reform. (2011). *IRS e-file and identity theft*. Retrieved July 24, 2012, from http://www.c-spanvideo.org/program/299848-1

IDManagement.gov. (2013). *Approved identity providers*. Retreived from http://www.idmanagement.gov/pages.cfm/page/icam-trustframework-idp, 2-8

Internal Revenue Service. (2008). *Advancing e-file study phase 1 report*. Washington, DC: IRS. Retrieved from http://www.irs.gov/pub/irs-utl/irs_advancing_e-file_study_phase_1_report_v1.3.pdf

Internal Revenue Service. (2011a). *The agency, its mission and statutory authority.* Washington, DC: IRS. Retrieved from http://www.irs.gov/uac/The-Agency,-its-Mission-and-Statutory-Authority.

Internal Revenue Service. (2011b). *National taxpayer advocate: 2011 annual report to Congress.* Washington, DC: IRS. Retrieved from http://www.irs.gov/pub/irs-utl/irs_tas_arc_2011_vol_1.pdf

Internal Revenue Service. (2012a). *Cost estimate reference FY 2011.* Department of the Treasury Document 6746 (Rev. 3-2012) Catalog Number 62707C. Washington, DC: IRS.

Internal Revenue Service. (2012b). *E-services—Online tools for tax professionals.* Washington, DC: IRS. Retrieved from http://www.irs.gov/taxpros/article/0,,id=109646,00.html

Internal Revenue Service. (2012c, March). *Internal Revenue Service data book, 2011.* Publication 55B. Washington, DC: IRS. Retrieved from http://www.irs.gov/pub/irs-soi/11databk.pdf

Internal Revenue Service. (2012d). *Taxpayer identification numbers (TIN).* Washington, DC: IRS. Retrieved from http://www.irs.gov/businesses/small/international/article/0,,id=96696,00.html

Internal Revenue Service. (2012e). *2012 and prior year filing season statistics.* Washington, DC: IRS. Retrieved from http://www.irs.gov/uac/2012-and-Prior-Year-Filing-Season-Statistics

IRS Online Services. (2012a, February). IRS online services—ACH debit business case. Washington, DC: IRS. Provided to RTI by IRS staff.

IRS Online Services. (2012b, January). IRS online service product management: Where's my amended return (WMAR) funding request. Provided to RTI by IRS staff. Washington, DC: IRS.

IRS Online Services. (2012c, March). IRS online services product management: Where's my refund FS 2013 funding request. Provided to RTI by IRS staff. Washington, DC: IRS.

IRS Online Services. (2012d, January). IRS online services product management: Transcript products business cases. Provided to RTI by IRS staff. Washington, DC: IRS.

IRS Online Services. (2012e, August). Taxpayer digital communication roadmap. Provided to RTI by IRS staff. Washington, DC: IRS.

iTrust. (2010). About NIH iTrust. National Institutes of Health, Washington, DC: iTrust. Retrieved from https://federation.nih.gov/itrust/aboutus.html

Javelin Strategy & Research. (2011, February). *2011 identity fraud survey report: Consumer version.*

Kitten, T. (2013, February 26). DDoS Attacks on Banks Resume. BankInfoSecurity. Retrieved from http://www.bankinfosecurity.com/ddos-attacks-on-banks-resume-a-5541

Kuhn, R. D., Hu, V. C., Polk, W. T., & Chang, S. (2011, February 26). *Introduction to public key technology and the federal PKI infrastructure.* Gaithersburg, MD: National Institute for Standards and Technology. Retrieved from http://csrc.nist.gov/publications/nistpubs/800-32/sp800-32.pdf

Libicki, M. C., Balkovich, E., Jackson, B. A., Rudavsky, R., & Watkins Webb, K. (2011). *Influences on the adoption of multifactor authentication.* Prepared for the National Institute for Standards and Technology. Retrieved from http://www.rand.org/content/dam/rand/pubs/technical_reports/2011/RAND_TR937.pdf

Malo, J. (2012, December 27). *TechSpend: Authentication and identity management technology spending is expected to grow 43% by 2016.* Corporate Executive Board blog post. Retrieved from http://www.executiveboard.com/towergroup-blog/techspend-authentication-and-identity-management-technology-spending-is-expected-to-grow-43-by-2016/

Miller, S. (2012, April 19). Written testimony of Steven T. Miller, Deputy Commissioner for Services and Enforcement Internal Revenue Service, before the House Committee on Oversight and Government Reform Subcommittee on Government Organization, Efficiency and Financial Management on the Tax Gap and Identity Theft. Retrieved from http://oversight.house.gov/wp-content/uploads/2012/04/4-19-12-Miller-Testimony.pdf

National Institute of Standards and Technology. (2006, March). Personal identity verification (PIV) of federal employees and contractors. Retrieved from http://csrc.nist.gov/publications/fips/fips201-1/FIPS-201-1-chng1.pdf

National Institute of Standards and Technology. (2012a). Recommendations for establishing an identity ecosystem governance structure. Retrieved from http://www.nist.gov/nstic/2012-nstic-governance-recs.pdf

National Institute of Standards and Technology. (2012b). Workshops: Previous workshops. Retrieved from http://www.nist.gov/nstic/workshops-home.html

Office of Management and Budget. (2011, October). Memorandum for chief information officers of executive departments and agencies: Requirements for accepting externally-issued identity credentials. Retrieved from http://www.cio.gov/Documents/OMBReqforAcceptingExternally_IssuedIdCred10-6-2011.pdf

Radicati, S., & Hoang, Q. (2012, April). *Email statistics report, 2012-2016.* Retrieved from http://www.radicati.com/wp/wp-content/uploads/2012/04/Email-Statistics-Report-2012-2016-Executive-Summary.pdf

Rogers, E. (1962). *Diffusion of innovations.* New York: Free Press of Glencoe.

Social Security Administration. (2012). Applying for disability benefits for myself. Retrieved from http://www.socialsecurity.gov/info/isba/disability/firstpartydib.htm

The White House. (2011, April). *National strategy for trusted identities in cyberspace: Enhancing online choice, efficiency, security, and privacy.* Retrieved from http://www.whitehouse.gov/sites/default/files/rss_viewer/NSTICstrategy_041511.pdf

Treasury Inspector General for Tax Administration (TIGTA). (2012). There are billions of dollars in undetected tax refund fraud resulting from identity theft. Retrieved from http://www.treasury.gov/tigta/auditreports/2012reports/201242080fr.html

U.S. Census Bureau (Census). (2012, November 16). *Quarterly Retail E-Commerce Sales: 3rd Quarter 2012.* Washington, DC. Retrieved from https://www.census.gov/retail/mrts/www/data/pdf/ec_current.pdf

U.S. Department of Agriculture. (2012). *Create an account.* Retrieved from http://www.eauth.egov.usda.gov/eauthCreateAccount.html

U.S. Department of Homeland Security. (2004). *Homeland Security presidential directive 12: Policy for a common identification standard for federal employees and contractors.* Retrieved from http://www.dhs.gov/xabout/laws/gc_1217616624097.shtm#0

U.S. Postal Service. (2012). Federal Cloud Credential Exchange (FCCX) request for proposals: Statement of objectives. Released on December 28, 2012. Retrieved from https://www.fbo.gov/index?tab=documents&tabmode=form&subtab=core&tabid=1ea07b c4ae987acb2be601e37bc3c17f

Verizon. (2012). *2012 data breach investigations report.* Prepared by the Verizon RISK Team. Retrieved from http://www.verizonbusiness.com/resources/reports/rp_data-breach-investigations-report-2012-ebk_en_xg.pdf

Weeks, R. (2003). Speak to concerns, not just to benefits. *iMedia Connection.* Retrieved from http://www.imediaconnection.com/content/2219.asp

Walker, M. B. (2012, October 22). CMS looks to NSTIC for identity management. Retrieved from http://www.fiercegovernmentit.com/story/cms-looks-nstic-identity-management/2012-10-22

Wisniewski, T., Nadalin, T., Cantor, S., Hodges, J., & Mishra, P. (2005, April 12). SAML V2.0 executive overview. *OASIS.* Retrieved from http://www.oasis-open.org/committees/download.php/13525/sstc-saml-exec-overview-2.0-cd-01-2col.pdf

# APPENDIX A:
# FICAM: ADDITIONAL INFORMATION

Federal government agencies all must comply with the FICAM Roadmap, which provides a unified approach for federal government agencies to manage identities of employees and of citizens who interact with federal agencies online. Released in 2009 and updated in 2011, FICAM primarily guides the implementation of standards that have already been mandated.[1] In particular, three specific federal initiatives require government agencies to adopt a common standard in the areas of risk assessment, credential issuing and managing, and encryption infrastructure:

- **E-authentication[2] Policy Framework** addresses a methodology for risk assessment and quantification of risk levels for electronic transactions.

- **PIV Initiative and Homeland Security Presidential Directive 12 (HSPD-12)** establishes the requirements of a centralized government credential.

- **Federal Public Key Infrastructure (PKI) Initiative** addresses the implementation of a centralized encryption infrastructure.

These three initiatives are considered to be the major drivers of FICAM, which acts to facilitate the compliance with and adoption of these and other initiatives across agencies. Additional detail on these three initiatives is provided below.

The FICAM framework mandates a centralized credential and identity management system to be used by the federal government. That is, under this framework, federal agencies must use a centralized system to manage credentials and identities (an example of this is the PIV card in use across the federal government). Applied to an agency such as the IRS, FICAM requires that agency employees and contractors use common government-sourced credentials (such as PIV cards) to access secure physical spaces and online systems.

In terms of communication and transactions with taxpayers, a specific aim of the U.S. government's efforts to standardize credentials is to improve the way that government agencies interact with citizens by reducing government costs and improving citizens' experience (e.g., through reducing time spent interacting with federal service providers, reducing identity theft, and providing additional and improved services). The FICAM Roadmap recommends that

---

[1] Throughout the remainder of this report, references to the FICAM Roadmap should be assumed to be referencing the 2011 document (FICAM version 2.0) unless otherwise stated.

[2] E-Authentication is short-hand for electronic authentication; however, the federal government uses the terms to refer to a government wide initiative to improve electronic authentication (as described in FICAM). Individual federal government agencies also commonly refer to their agency-specific online authentication system or initiatives by the name E-Authentication. In this report, the term E-Authentication will be used both in reference to the federal government-wide initiative (as here) as well as in reference to electronic authentication efforts ongoing at the IRS and other individual government agencies.

federal agencies seek to accept third-party credentials, and it provides a set of standards, process requirements, and decision support tools. The Roadmap both guides and in some cases mandates that federal government agencies engage in certain activities related to the development of authentication systems that can accept third-party credentials. For example, the Roadmap discusses the requirement that federal government agencies must select identity providers who have been certified by OMB as able to provide a certain LOA and able to comply with federal government standards and standard protocols.

Additional information is provided in the FICAM Roadmap to help support agencies' decision making processes. The following recommended steps are listed for agencies considering accepting third-party credentials: determine the required LOA for a particular application, identify the required/preferred credentials that meet the LOA, analyze the user population to assess willingness to adopt, and finalize the credential requirements.

The FICAM Roadmap document demonstrates its vision of third-party credential acceptance by presenting a scenario describing a citizen using a third-party credential to access a federal research website. In this scenario, the user is prompted to provide a third-party credential to the RP (i.e., the research website). The user would be linked to the site of the identity provider and would provide a username and password to the provider. Upon successful authentication with the identity provider, the system would send an assertion to the RP, showing that the user's credentials are valid.

For federal government agencies, FICAM offers technical guidance for establishing a common, agency-independent management approach for identities, credentials, and access control. However, FICAM does not give specific technical guidance on how to implement the infrastructure needed at a specific agency. Discussing the gap in acceptance of third-party credentials, the FICAM Roadmap states, "[t]he Federal Government needs federation processes such as direct relationships with trusted Identity Providers, working with Trust Broker services, or by entering into a federation of trust" (Federal CIO Council, 2011, p. 128).

Broadly, in terms of third-party authentication, the following needs exist:

- standard agreements/standard operating procedures concerning stakeholder relationships (e.g., IDPs and RPs): *"what should the pricing structure (e.g., cost per transaction or cost per user) for transactions between and IDPs and an RP be??"*

- process for certifying/accrediting stakeholders providing identity proofing and authentication: *"how can RPs trust IDPs?"*

- standards, protocols, and policies dictating how and what information will be exchanged between stakeholders: *"how should IDPs and RPs interact/communicate?"*

Table A-1 provides a list of the gaps that were identified by FICAM (2011) as critical for the FICAM goals to be implemented. They include a need for standards, standard architectures, and interoperability frameworks. Many of these gaps still exist and are relevant to the adoption of NSTIC-aligned third-party credentials.

**Table A-1.    Gaps and Recommendations for FICAM Framework Implementation**

| Item | Performance Gap | Performance Improvement Recommendation |
|---|---|---|
| 1 | No common definition or data specification identifying the minimum data elements for creating and sharing digital identity data. | Develop and implement a government-wide digital identity data specification to standardize and streamline collection, management, and sharing of identity data for an individual. |
| 2 | Need for common definitions of additional identity attributes required for mission-specific functions. | Implement Backend Attribute Exchange (BAE) common data elements or other shared attribute exchange models to support data sharing of common, mission-specific identity attributes outside of the digital identity data elements within specific communities of interest. |
| 3 | Inability to correlate and synchronize digital identity records and automatically push and pull identity data between systems. | Develop an Authoritative Attribute Exchange Service at the agency level to index and link authoritative sources of identity data and synchronize digital identity records for an individual. |
| 4 | Lack of authoritative sources for contractor/affiliate identity data. | Establish a government-wide approach for creating and maintaining contractor and affiliate identity data that can be used across agencies. |
| 5 | Prevalence of redundant collection and management of digital identity data for the same user. | Modify processes and systems such that identity data may be collected once and linked to authoritative sources throughout the enterprise for management and use of the data. |
| 6 | Need for a capability to bind externally issued credentials to an agency's identity record for an external user. | Develop and implement approaches and technologies enabling the linking of third-party credentials to the digital identity records of external users for use in application access. |
| 7 | Lack of reciprocity in the acceptance of background investigations completed by or on behalf of another agency. | Resolve process and technology shortfalls preventing agencies from referencing and honoring reciprocity of background investigations for individuals adjudicated by another agency. |
| 8 | Lack of integration between PIV enrollment and background investigation processes. | Close process gap to ensure that the fingerprints used in processing background investigations are collected as part of the PIV enrollment process and submitted electronically. |
| 9 | No capability to reference prior background investigation for an individual based on fingerprint biometric. | Establish capability to tie an individual to a prior background investigation based on referencing fingerprints. |

(continued)

**Table A-1.    Gaps and Recommendations for FICAM Framework Implementation (continued)**

| Item | Performance Gap | Performance Improvement Recommendation |
|------|-----------------|----------------------------------------|
| 10 | Lack of integration between PIV systems and FEMA Emergency Response Official Repository. | Integrate PIV systems with F/ERO database to provide required data. |
| 11 | Redundant credentialing processes. | Reduce the number of credentials issued for the same individual within and across agencies and enable the use of PIV and other credentials that have already been issued. |
| 12 | Underutilization of PIV certificates as primary PKI credentials for internal users. | Enable the use of PIV certificates across the enterprise and eliminate redundant credentials. |
| 13 | Lack of government-wide approach and guidance for managing key history. | Provide guidance on the management of key history. |
| 14 | Lack of product adoption for path discovery and validation. | Implement path discovery and validation products. |
| 15 | Administrative and user burden associated with managing and remembering numerous federally issued stand-alone password tokens. | Minimize the reliance on password tokens by enabling PIV card usage for internal users and the acceptance of externally issued credentials for external users. |
| 16 | Lack of automation in provisioning workflows. | Implement automated processes and technologies to provision or deprovision users based on established business rules. Eliminate manual provisioning processes by tying applications/systems into the automated workflow. |
| 17 | Inability to perform cross-agency provisioning. | Work collaboratively to establish business rules for sharing identity/access record data as needed between agencies in order to provision access. |
| 18 | Lack of government-wide approach for provisioning logical access for external users. | Work collaboratively to determine approach for provisioning logical access for external users at all assurance levels. |
| 19 | Inability of many installed PACS technologies to meet new requirements for electronic authentication outlined in SP 800-116. | Upgrade current processes and technologies to meet requirements. |
| 20 | Lack of integration between PACS and other ICAM systems (provisioning and credentialing systems). | Federate PACS with other ICAM systems to allow sharing of user attributes and credential information from authoritative sources. |
| 21 | Lack of automation and consistency in agency processes/systems used for visitor access control. | Upgrade technologies to support secure, automated processes for requesting and provisioning visitor access. |
| 22 | Inability to electronically authenticate and accept PIV and PIV-interoperable (PIV-I) credentials from visitors. | Enable the use of PIV and PIV-I cards for visitor access. |

(continued)

**Table A-1.    Gaps and Recommendations for FICAM Framework Implementation (continued)**

| Item | Performance Gap | Performance Improvement Recommendation |
|---|---|---|
| 23 | Need for enterprise-wide access management capability at the agency level. | Implement processes and technologies to support an agency-wide approach for managing logical access that links individual applications to a common access management infrastructure wherever possible. |
| 24 | Insufficient maturity in BAE implementation to support cross-agency data exchange in access scenarios. | Provide implementation guidance based on pilot deployment of the BAE to further enable ability to share data across agencies. |
| 25 | Lack of government-wide guidance regarding use of encryption and digital signatures. | Develop government-wide implementation guidance for the use of encryption and digital signatures. |
| 26 | Lack of adoption of PKI technologies and processes. | Fully enable the use of the PIV credential to further encryption and digital signature usage. |

Source: Federal CIO Council (2011).

# APPENDIX B:
# ADDITIONAL APPROACHES TO IMPROVE IRS AUTHENTICATION

In addition to the electronic authentication mechanism discussed in the main report, the IRS could make several other changes to improve authentication. For example, the IRS could extend PINs to all taxpayers, add more security questions as part of the online and offline authentication process for taxpayers, and streamline income verification. Although it would be difficult to achieve the same level of security as with FICAM or through proprietary means, authentication would be improved. Of note, the items were identified based on both a literature review and interviews with IRS staff; all are being considered and some may even be in the process of being adopted at the date of this report's publication.

## B.1    IP PIN and Additional Security Questions

One security measure the IRS provides is IP PINs, which are PINs given to taxpayers confirmed as victims of identity theft. In the 2012 filing season, 252,000 victims and at-risk taxpayers were given IP PINs,[1] which they are required to provide when filing returns in the following year. The IP PIN acts as an additional layer of security that supplements current methods of authentication, rather than replacing them. Although this step adds to the complexity of securely filing a return, it adds an assurance of security for those who feel especially at risk of identity theft.

A potential authentication solution would involve expanding the IP PIN program to all taxpayers rather than only those that are at special risk. This would add a layer of security for all of the approximately 150 million taxpayers. Although the IP PIN would give each taxpayer one more credential to safeguard, it could help reduce the incidence of identity theft.

In addition to the IP PINs, the IRS may supplement current authentication methods by presenting the taxpayer with security questions. These questions would be either previously provided by the taxpayer or would come from a third party. When signing a tax return, a taxpayer would have to answer one or more security questions to prove his or her identity. This could be used in concert with an IP PIN or as an alternative to it, depending on the usability of an implementation.

There would be several ways of collecting answers to the security questions:

- Establish a registration process in which a taxpayer is prompted with questions on the IRS website and provides answers that are stored by the IRS. This would require a new, secure online service.

---

[1] See http://waysandmeans.house.gov/UploadedFiles/George_Testimony.pdf, p. 3.

- Establish a relationship with a credit bureau and develop a protocol for the bureau to verify taxpayers' identities by asking a series of questions (e.g., relating to the taxpayer's credit history or past addresses) and assert the identity to the IRS. This would require a new online service to redirect the taxpayer to a credit bureau web page where he or she would be prompted with the security questions.

- Collect secret question/answer combinations from credit bureaus. The questions collected by the IRS would then be presented to the taxpayer upon filing of a tax return, without the need for redirecting the taxpayer to a third party or collecting the information from the taxpayer beforehand.

The extension of the IP PIN program to all taxpayers and the addition of security questions/answers would give taxpayers a stronger assurance of security when they file. This would increase adoption of e-filing, however, only if the additional security controls are easy to use. If taxpayers become frustrated or confused during the process of enrollment or identity verification, they may opt to file a paper return instead.

## B.2    Streamlined Income Verification

A subset of taxpayer returns also go through an income verification process. This process compares self-reported income from a tax return with income data from forms filed by third parties such as employers and financial institutions. For instance, if a taxpayer claims $50,000 in income received from her employer, the IRS may compare this number with that on the W2 form provided by her employer. However, third-party income verification has several disadvantages. Receipt of information returns is often too late, and verification of income data is often too slow.

While an employer must issue a W2 or a 1099 form to the taxpayer no later than January 31, a copy of the form must also be sent to the IRS (in the case of 1099 forms) and to the Social Security Administration (in the case of W2 forms) by March 31 (IRS, 2011b). The IRS receives 1099 forms as they arrive, and W2 forms are transmitted weekly from the SSA. However, because of the differences in deadlines, there is no guarantee that third-party forms will be on file when a taxpayer has already submitted a tax return.

Moreover, when the IRS does have the third-party income forms along with the tax return for a particular taxpayer, verifying the forms can be very slow. As described during an interview with IRS staff, the IRS recently adopted a new automated tool to speed up income verification; however, because of a lack of sufficient technical infrastructure, some verification of income is still conducted manually. This results in wait times as long as several weeks from the receipt of the necessary documents to the successful comparison of income data. Because of the delays in receiving and verifying third-party income data, the IRS does not withhold a tax refund before verifying income; thus, many returns are unverified or are verified after a refund has already been issued to the taxpayer.

There are three solutions to expedite the verification of income data:

- **Require employers to submit tax forms earlier:** To ensure that third-party income forms are received at the same time tax returns are received, the policy would need to be changed to mandate a single deadline for employers to submit tax forms to employees and to the federal government. That is, employers and other third parties should be mandated to send W2 and 1099 forms to the federal government as soon as forms are sent to employees.

- **Expedite the transmission of W2 forms from the SSA to the IRS:** Because W2 forms are transmitted from the SSA weekly, these forms should be sent faster so the IRS can have more real-time access to the data. Changing policy and improving data transmission from the SSA would ensure that taxpayer-reported data and third-party reported data are received as close together as possible.

- **Automate the comparison of available income data:** The IRS could make the necessary investments to establish the resources and capability to expediently compare income data on a return with data provided from an employer. This would require a significant investment in computing power and software to efficiently compare the up to 150 million returns with their associated third-party forms. Furthermore, costs will vary depending on the format that third-party data are transmitted in. Employers may send W2 and 1099 forms on paper or electronically.

Income verification can be applied to the vast majority of return filers, and it can minimize many types of identity theft. Although some filers may claim no income or have income that cannot be corroborated by a third party, the returns can be more easily scrutinized while the more common returns with verifiable income are automatically checked. Moreover, this step would enable the IRS to effectively protect revenue beyond identity fraud, including taxpayers who purposefully underreport income.

## B.3  Costs of New Authentication

The traditional solution has two distinct implementations. The extended IP PIN program and security questions can be developed together in a cost-efficient way, so they may be packaged as a single solution. The automated income verification program is a separate solution and may be used as an alternative approach. Tables B-1 and B-2 provide a list of the costs involved.

Approach A calls for an upgrade of the IRS's infrastructure to enable the automatic comparison of income data provided by employers and with that provided by the taxpayer. Approach B is a more general cost, calling for a policy change to mandate employers to submit all forms by January 31. The costs of this approach are not clear and are complicated by various factors both within and beyond the IRS's control.

**Table B-1.    Costs of an Extended IP PIN and Security Questions Program**

| Program | Type of Cost | Activity |
|---|---|---|
| Extending IP PIN | Up-front cost | ▪ Upgrading infrastructure to support the extension of the IP PIN program from 250,000 users to approximately 150 million users<br>▪ Mailing out IP PIN notices (typically two notices for each taxpayer) to all taxpayers |
| | Ongoing cost | ▪ Supporting taxpayers who have lost or forgotten their IP PINs |
| Security questions | Up-front cost | ▪ *Approach A:* Establishing a registration process to collect secret questions/answers from taxpayers<br>▪ *Approach B:* Establishing a service agreement and protocol with a credit bureau to provide secret questions/answers for all taxpayers<br>▪ *Approach C:* Establishing a service agreement and protocol with a credit bureau to conduct its own authentication and assert each taxpayer's identity back to the IRS<br>▪ *Approaches A/B:* Upgrading infrastructure to store secret questions/answers for each taxpayer, provide interface, and verify entered data |
| | Ongoing cost | ▪ *Approach A/B:* Supporting taxpayers who have forgotten answers to secret questions<br>▪ *Approach C:* Per-transaction costs for authentication by credit bureau |

**Table B-2.    Costs of an Automated Income Verification Program**

| Program | Type of Cost | Activity |
|---|---|---|
| Automated Income Verification | Up-front cost | ▪ *Approach A:* Upgrading infrastructure to automate the comparison of employer-provided forms with tax returns<br>▪ *Approach B:* Mandating employers to send W2 and 1099 forms to the IRS by January 31 |
| | Ongoing cost | ▪ *Approach A:* Manually checking returns that failed automatic screening |

**B-4**

# APPENDIX C:
# DATA USED FOR ANALYSIS CALCULATIONS

The data used for the final analysis calculations, the results of which are summarized in Chapter 5, were acquired through interviews with IRS staff and identity experts, internal IRS documents provided to RTI, and public documents. Tables C-1 and C-2 list the primary data used for this analysis and the sources.

**Table C-1.    Data Used for Calculating Savings from Online Services**

| Category of Cost-Saving | Metric | Value | Source |
|---|---|---|---|
| Online transcripts view-and-print capability (eTranscripts Project Phase 2) | Annual number of calls requesting a transcript | 3,400,000 | IRS (2012e) |
| | x Average cost per call | $30.00 | |
| | *Total call center cost* | *$102,000,000* | |
| | Annual number of walk-in requests at a Taxpayer Assistance Center (TAC) | 350,000 | |
| | x Average cost per walk-in request | $12.50 | |
| | *Total walk-in cost* | *$4,357,500* | |
| | Annual number of requests using form 4506-T[a] | 4,400,000 | |
| | x Average cost per form 4506-T request | $2.30 | |
| | *Total form 4506-T cost* | *$10,296,000* | |
| | Annual number of requests using Order a Transcript online tool | 1,800,000 | |
| | x Average cost per request with online tool | $1.00 | |
| | *Total online tool request cost* | *$1,800,000* | |
| | **Total potential cost reduction** | **$118,453,500** | |
| | **Up-front investment cost** | **$2,803,333** | |
| | **Annual recurring cost** | **560,667** | |
| Where's My Refund | Annual number of calls checking the status of a refund | 7,000,000 | IRS (2012c) |
| | x Average cost per call | $30.00 | |
| | **Total potential cost reduction** | **$210,070,000** | |
| | **Up-front investment cost** | **$5,860,000** | |
| | **Annual recurring cost** | **$1,172,000** | |
| Where's My Amended Return | Annual number of calls from taxpayers checking on the status of an amended return | 763,000 | IRS (2012b) |
| | × Average cost per call | $30.00 | |
| | **Total potential cost reduction** | **$22,897,630** | |
| | **Up-front investment cost** | **$2,050,000** | |
| | **Annual recurring cost** | **$410,000** | |

(continued)

**Table C-1.    Data Used for Calculating Savings from Online Services (continued)**

| Category of Cost-Saving | Metric | Value | Source |
|---|---|---|---|
| ACH debit | Annual cost of paper checks received | $72,800,000 | IRS (2012a) |
| | + Annual lost interest on paper money payments based on delay | $30,900,000 | |
| | **Total potential cost savings** | **$103,700,000** | |
| | **Up-front investment cost** | **$500,000** | |
| | **Annual recurring cost** | **$100,000** | |
| Secure webmail portal (for transmission of notices with PII) | Annual number of notices sent that do not need to be sent on paper | 106,800,000 | IRS (2012d) and interview with IRS staff |
| | × Cost per notice sent | $3.29 | |
| | *Potential cost savings from notices sent* | *$351,372,000* | |
| | Average minutes saved by case worker from e-mail vs. phone call | 5.0 | |
| | Estimated annual number of phone calls that could be switched to e-mail | 10,500,000 | |
| | Total FTEs saved | 583.3 | |
| | Cost per FTE | $60,000 | |
| | *Potential cost savings from reduced calls received/placed* | *$35,000,000* | |
| | **Total potential cost reduction** | **$386,372,000** | |
| | **Up-front investment cost** | **$2,803,333[b]** | |
| | **Annual recurring cost** | **$560,667[b]** | |
| Manual notification tool | Annual number of cases worked by IRS staff | 2,500,000 | Interview with IRS |
| | Approximate number of contacts per case | 1.6 | |
| | Estimated annual number of non-PII notifications that could be sent by e-mail | 4,000,000 | |
| | Average minutes saved by case worker from e-mail vs. phone call | 5.0 | |
| | Total FTEs saved | 139 | |
| | Cost per FTE | $60,000 | |
| | **Total potential cost reduction** | **$8,333,333** | |
| | **Up-front investment cost** | **$2,803,333[b]** | |
| | **Annual recurring cost** | **$560,667[b]** | |

(continued)

**Table C-1.    Data Used for Calculating Savings from Online Services (continued)**

| New Online Service | Metric | Value | Source |
|---|---|---|---|
| Automated digital notification tool | Annual number of non-PII notices sent | 23,140,000 | Interviews with IRS |
| | +Annual number of CP521 notices sent | 29,000,000 | |
| | Total number of notices sent per year | 51,140,000 | |
| | × Postage cost per notice sent | $3.29 | |
| | **Total potential cost reduction** | **$171,540,600.00** | |
| | **Up-front investment cost** | **$2,803,333**[b] | |
| | **Annual recurring cost** | **$560,667**[b] | |
| Live chat | Estimated annual number of hours spent on nonaccount calls that could be addressed in live chat | 6,400,000 | Interview with IRS |
| | Productivity gain from live chat | 30.0% | |
| | Total FTEs saved from productivity gain | 1,280 | |
| | Cost per FTE | $60,000 | |
| | *Total potential savings from nonaccount calls* | **$76,800,000** | |
| | Estimated annual number of hours spent on *account-related* calls that could be addressed in live chat | 7,100,000 | |
| | Productivity gain from live chat | 30.0% | |
| | Total FTEs saved from productivity gain | 1,420 | |
| | Cost per FTE | $60,000 | |
| | *Total potential savings from account-related calls* | *$85,200,000* | |
| | **Total potential cost reduction** | **$162,000,000** | |
| | **Up-front investment cost** | **$2,803,333**[b] | |
| | **Annual recurring cost** | **$560,667**[b] | |
| Document transfer | Annual number of cases worked by IRS staff | 2,500,000 | Interview with IRS |
| | × Average number of documents received per case | 0.9 | |
| | Total number of documents handled in case work | 2,300,000 | |
| | Time saved by case worker (minutes) | 5 | |
| | Total FTEs saved | 127 | |
| | Cost per FTE | $60,000 | |
| | **Total potential cost reduction** | **$8,000,000** | |
| | **Up-front investment cost** | **$2,803,333**[b] | |
| | **Annual recurring cost** | **$560,667**[b] | |

[a]. Form 4506-T is a Request for Transcript of Tax Return. This form is used to order a transcript or other return information free of charge from the IRS.

[b]. These cost estimates for the up-front investment costs and the recurring costs were estimated based on taking the average of the up-front investment costs and the recurring costs, respectively, for the other applications for which costs estimated were provided by the IRS.

**Table C-2.** **Data Used for Calculating Savings from Reduced Authentication Costs**

| Current Authentication Activity | Metric | Value | Source |
|---|---|---|---|
| Registered user portal (RUP) customer service requests | Number of calls regarding credential issues | 93,000 | Interview with IRS |
| | × Minutes per call | 12.5 | |
| | ÷ Minutes per hour | 60 | |
| | Total hours spent answering calls for credential issues | 19,375 | |
| | ÷ Hours per FTE per year | 2,080 | |
| | Total FTEs spent handling calls | 9.3 | |
| | × Cost per FTE | $72,800 | |
| | **Total potential cost reduction** | **$678,125** | |
| E-file PIN customer service requests | Number of assistor-answered calls for e-file PIN support | 225,406 | Interview with IRS |
| | × Cost per e-file PIN support call | $9.33 | |
| | **Total potential cost reduction** | **$2,103,038** | |

**Table C-3.** **Data Used for Calculating OLS Adoption Costs**

| Online Service | Adoption Cost |
|---|---|
| eTranscripts Phase II | $2,803 |
| ACH Debit | $500 |
| Where's My Refund | $5,860 |
| Where's My Amended Return | $2,050 |
| Secure Webmail Portal | $2,803 |
| Manual Notification Tool | $2,803 |
| Automated Digital Notification Tool | $2,803 |
| Live Chat | $2,803 |
| Document Transfer | $2,803 |
| **TOTAL** | **$25,230** |

**Table C-4.    Data Used for Calculating OLS Annual O&M Costs**

| Online Service | O&M Cost |
|---|---|
| eTranscripts Phase II | $561 |
| ACH Debit | $100 |
| Where's My Refund | $1,172 |
| Where's My Amended Return | $410 |
| Secure Webmail Portal | $561 |
| Manual Notification Tool | $561 |
| Automated Digital Notification Tool | $561 |
| Live Chat | $561 |
| Document Transfer | $561 |
| **TOTAL** | **$5,046** |

www.ingramcontent.com/pod-product-compliance
Lightning Source LLC
Chambersburg PA
CBHW081547170526
45166CB00009B/2611